D1483560

CALIFORNIA NATURAL HISTORY GUIDES

RAPTORS OF CALIFORNIA

California Natural History Guides

Phyllis M. Faber and Bruce M. Pavlik, General Editors

RAPTORS
of California

Hans Peeters
Pam Peeters

Illustrations and photos by Hans Peeters

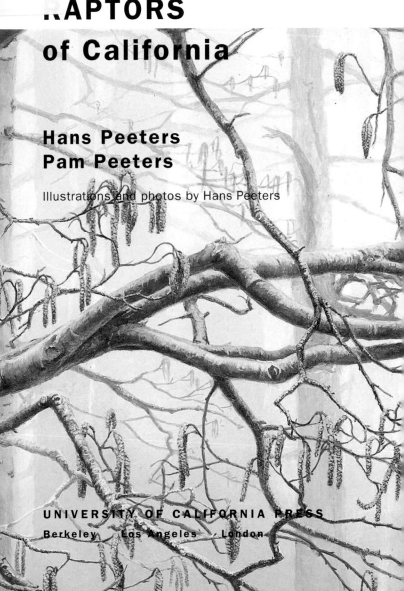

UNIVERSITY OF CALIFORNIA PRESS

Berkeley   Los Angeles   London

Dedicated to Tom Cade and Bob Risebrough, each of whom
has, in his own way, done so much for raptor restoration
and conservation.

California Natural History Guide Series No. 82

University of California Press
Berkeley and Los Angeles, California

University of California Press, Ltd.
London, England

© 2005 by the Regents of the University of California

Library of Congress Cataloging-in-Publication Data

Peeters, Hans, 1937–.
    Raptors of California / Hans Peeters, Pam Peeters; illustrations and
photographs by Hans Peeters.
        p.   cm.—(California natural history guides ; 82)
    Includes bibliographical references and index.
    ISBN 0-520-23708-0 (cloth : alk. paper)—ISBN 0-520-24200-9 (pbk. : alk.
paper)
    1. Birds of prey—California.   2. Birds of prey—California—Identification.
I. Peeters, Pam, 1949–   II. Title.   III. Series.
QL677.78.P44   2005
598.9'09794—dc22

                                                                                    2004006946

Manufactured in China
10   09   08   07   06   05
10   9   8   7   6   5   4   3   2   1

Cover: Anatum Peregrine *(Falco peregrinus anatum),* painting by Hans Peeters.

The publisher gratefully acknowledges the generous
contributions to this book provided by

the Moore Family Foundation
Richard & Rhoda Goldman Fund
and
the General Endowment Fund of the
University of California Press Associates.

CONTENTS

Plates follow page 212

PREFACE

Several features set this field guide apart from others. Rather than focusing on the valuable but mind-numbing minutiae of points of identification, we have added a new dimension to finding, identifying, and watching raptors in the field by providing the observer with an understanding of each species and its behavior and also of raptors as a group. Considerable space is devoted to foraging habits, for example, because hawks spend much time looking for prey and are often observed doing so. Some facets of raptor and general bird biology have been included because they add enjoyment to raptor watching and can help with species recognition. Personality traits of some species are outlined to facilitate interpretation of behavior and perhaps suggest the approachability of a raptor. To help biologists and Breeding Bird Atlas volunteers gauge breeding population numbers, a section on how to find nests is included; and basic breeding information is supplied so that if you find a hawk's nest, you will know how long to expect the bird to sit on it (provided of course that it was found at the onset of incubation), how long it takes for the hatched young to grow juvenal plumage (so you can identify the species if you have not seen the parents), and how to interpret the most likely displays of the adults. Frequently, voice and behavior draw the viewer's eye and thereby help in spotting raptors; a description of the most likely heard vocalizations is therefore included.

References for specific information have been cited in the text selectively, principally to help the reader find sources for research that is new or not widely reported. Where information is provided without a citation, it is often based on our own unpublished observations from a lifetime of watching hawks in California (and elsewhere), or it has been widely published in raptor literature.

Although a substantial number of California raptor species occur throughout North America, eastern populations may differ in subtle ways from those of the west. For example, eastern Red-tailed Hawks *(Buteo jamaicensis),* in addition to being somewhat different in color (they form another race), also have shorter tails than do ours, and the eastern Cooper's Hawk *(Accipiter cooperii)* is larger than the California form and so is not as easily confused with a Sharp-shinned Hawk *(A. striatus).*

The breeding distribution of many raptor species is highly dynamic; not only have Peregrine Falcons *(Falco peregrinus)* established (or reestablished) a presence where they were unknown a decade ago, but also the Red-shouldered Hawk *(Buteo lineatus)* has dramatically increased its breeding range in California in recent years, moving into previously unused areas. The range maps in this book, supported by our own knowledge and that of raptor biologists throughout the state, are perhaps the most accurate, up-to-date maps yet published for California's raptors. We believe, however, that although range maps are a common and popular feature of field guides, they are of limited use because distributions of birds of prey are often temporary, and maps can become quickly outdated. We mention some specific and general areas that have been productive for hawk-watching in the past and where there is a likelihood of seeing a certain species; however, it is more useful to direct hawk-watchers to habitats where specific raptors are likely to be seen. See the descriptions of various habitats at the beginning of the Species Accounts section.

A great many people contributed in one way or another to the writing of this book, and to all of them we extend our sincere thanks. The senior author owes his profound gratitude to Alfred Beckers of Cologne, Germany, who, as a young man, took a 10-year-old boy under his wing and introduced him to the wonders of hawks, the genesis of a friendship that has lasted for over half a century. Upon arriving in 1956 in Berkeley as a still slim student, he then had the good fortune of striking up what would prove to be lifelong friendships with three ardent raptorphiles, Sterling Bunnell, Steve Herman, and E.W. Jameson Jr. Soon our small group expanded to include Ed Hobbs, Grainger Hunt, Louis Davis, and as the years passed, Robert Risebrough, John Schmitt, and Brian Walton—hawk lovers all, who in one way or another shared ideas and experiences. Bruce Mahall spent countless, often memorable, hours in the field with us looking for raptor nests and exploring California.

We especially wish to thank Allen Fish and Lloyd Kiff for their careful reading of the first draft of this book and their helpful suggestions and gentle corrections; Howard Cogswell very generously provided some unpublished migrant raptor data from southern California.

Several of the experts we consulted, especially concerning breeding ranges, were known to us only by name and reputation; all were unfailingly generous with their time and cheerfully shared their knowledge. We are much indebted to the friends, acquaintances, and strangers who have aided us, whether by providing information or granting access to lands. These helpers include Dick Anderson, Eric Ariyoshi, Larry Baines, Doug Bell, Dave Bittner, Pete Bloom, Jeremy Bradshaw, Rose Britton, Seth Bunnell, Joe Burnett, Tom Cade, Doug Carmean, John Coleman, Joel Cracraft, Jim Dawson, Phil Detrich, Joe DiDonato, Tery Drager, Jamey Eddy, Art Edwards, Leon Elam, Sheila Foster, Amy Fesnock, David Garcelon, Phil Gordon, Jim Hatchett, Buzz Hull, Terry Hunt, Kevin Hunting, Mark Jenkins, Ron Jurek, John Keane, Tim Koopmann, Brian Latta, Janet Linthicum, Sal Lucido, Michael Mace, Steve Margolin, Frank Marino, Geoff Monk, Joe Naras, Bruce Palmer, Monica Parisi, Adrian Reuter, Ron Schlorff, Kelly Sorenson, Philip Unitt, Samuel Vasquez, Michael Wallace, Clayton White, and Brian Woodbridge.

Lastly, a special thanks must be reserved for our editors at University of California Press. Doris Kretchmer's enthusiasm, encouragement, and gentle prodding supplied the impetus to finish the manuscript. Scott Norton proved to be a fountain of good ideas and a master of tactful negotiation who, with unflagging good humor, fashioned our text and pictures into a book.

Map 1.
Basic California
Topography

Map 2.
California Counties

RAPTORS ARE AN integral part of the California landscape. Not only do they play major roles in various ecosystems and act as indicators of the ecological health of the state, but they are also important aesthetic components of our land; it is very difficult to imagine a typical California savanna of stately oaks without a Red-tailed Hawk *(Buteo jamaicensis)* soaring above it in lazy circles, nor would the jagged cliffs of Big Sur have the same magic without the scream of the gulls or the arrow shape of a Peregrine Falcon *(Falco peregrinus)* flashing across the crags.

California surely is one of the finest hawk-viewing areas in the United States. Although Texas and Arizona boast more species because they attract a few more subtropical raptors, these are very local; you can drive vast distances and see nary a hawk. In California, you can count on catching glimpses of American Kestrels *(Falco sparverius)* and Red-tailed Hawks nearly everywhere, in part because our generally mild climate provides for a year-round abundance and diversity of food for both raptors and prey.

In addition, California has a greater variety of habitats and climatic zones than any other state, which invite both specialized raptors and generalists. Because of the close proximity of these varied regions to each other, many species can be found with rel-

Fig. 1. California savanna of chiefly nonnative grasses and valley oaks *(Quercus lobata)*.

Fig. 2. Raptor food includes bobcats, crickets, lizards, other raptors, fish, dragonflies, centipedes, washed-up whales, worms, and maggots. A Turkey Vulture may eat pumpkins when better food is scarce.

atively little traveling. The high visibility of hawks in California also results from its position on the Pacific Flyway (and therefore from the influx and passage of winter migrants from the north), from its frequently open vistas, and from the presence of many hill ranges and mountains that encourage soaring.

The great Central Valley, with its wide-open expanses, is prime hawk-watching country; even on the densely forested heights of the Sierra one can locate raptors, although they are harder to find there. The coastal terraces, hills, and valleys in central and northern California, with their dense, low vegetation, harbor countless rodents, which make these areas magnets for wintering hawks; here, Rough-legged Hawks *(Buteo lagopus)* and Red-tailed Hawks of nearly all the color morphs join the resident Peregrine Falcons, hanging nearly motionless in the onshore wind. The white-water rivers of the north are fringed with conifers where the observer can easily spot the nest of an Osprey *(Pandion haliaeetus)* and, in some places, that of a Bald Eagle *(Haliaeetus leucocephalus)*. To the northeast and abutting against the eastern Sierra southward, the Great Basin spreads its sage plains into our state, along with isolated forest-covered ridges

Fig. 3. California habitats: northern coastal terraces are attractive to many raptors, especially in winter.

Fig. 4. California habitats: Great Basin and distant eastern Sierra Nevada.

and rolling hills, and here it is not difficult to watch Prairie Falcons *(Falco mexicanus)* and Golden Eagles *(Aquila chrysaetos).* Even our cities have become virtual raptor sanctuaries, with for-

Fig. 5. California habitats: Mojave Desert, shown here with Joshua trees *(Yucca brevifolia),* which are among the few available nest sites for raptors.

merly rarely seen or shy species such as Peregrine Falcons and Cooper's Hawks *(Accipiter cooperii)* nesting on high-rises and above residential streets. A few hardy raptor species actually thrive in the driest land in California, the deserts.

What Is a Raptor?

In common usage, the word "hawk" is an imprecise term that has been used interchangeably with "raptor" or "bird of prey" for centuries. The term "raptor" conjures up an image of a bird that is powerful and swift, qualities much admired by humans. At rest, a raptor sits very upright and gazes about calmly, exuding an air of what we see as self-confidence and nobility. However, the term itself is nonspecific and applies to very small and delicate hawks as well.

Technically, most of California's diurnal (day-active) raptors are called "true" hawks, with the eagles being the largest species; falcons, however, are placed in a separate family, as are American vultures. They all, nevertheless, feed chiefly on meat and have a beak that is hooked to facilitate tearing apart this food. In most, the feet bear sharp, strongly curved claws for seizing prey. Most are raptors, that is, they capture their own prey; but so are owls, which are not included in this book. Owls are considered nocturnal raptors, although some are active during the day.

The terms "hawk" and "raptor" have been used freely and interchangeably in this book to include all diurnal birds of prey, and although purists may object to such generalized usage, the alternatives are pointlessly cumbersome. Vultures, which are often lumped with the diurnal raptors, are included because they have many superficial characteristics similar to those of more typical raptors and, at a distance, often are difficult to tell apart from them.

The evolution of hawks is roughly comparable to that of motor vehicles. From a generalized ancestor, the motorized horse buggy, specialized vehicles of different shapes and sizes were developed for a variety of purposes. Although not all raptors may be derived from the same ancestor, most appear to be, and their diversification is similar. The Sharp-shinned Hawk (*Accipiter striatus*) is like a sports car: it is tiny and agile but has terrible fuel economy. The high metabolic rate demanded by its small size requires the intake of lots of food frequently, and a few days' bad luck in hunting can quickly lead to death. Its distant cousin, the Golden Eagle, is the 18-wheeler, ponderous and huge but very fuel efficient. In between these extremes lie the various hawks and falcons—the pickups, vans, sedans, and race cars of the auto world—which show variations in shape and size that adapt them to available habitats and prey.

All California vultures are members of the Cathartidae, the family of New World vultures. As raptors go, our vultures barely fall under the definition; they do so chiefly on the basis of their eaglelike appearance in flight, their hooked beaks, and to some extent their food.

Kites, hawks, harriers, ospreys, and eagles are raptors that have much in common, and although they may vary dramatically in size and shape, relating to their manner of foraging and the size and kind of their prey, all are usually pursuers of live quarry and are equipped with strong, grasping feet. All are in the family Accipitridae.

Falcons are the most streamlined birds of prey, with long, pointed wings and compact bodies; some are capable of attaining astonishing speeds. Falcons are members of the family Falconidae.

Raptor Names

The first English settlers on this continent were not well versed in natural history, which was not much in vogue then; they therefore tended to label organisms with the names of animals familiar to them from the old country, notwithstanding that the new birds and mammals they were seeing were in fact often not very similar to those of Merrie Olde England. To this day, Turkey Vultures *(Cathartes aura)* often are called "buzzards," wrongly, because to unobservant early settlers, they appeared similar to a true hawk, the Common Buzzard *(Buteo buteo),* of Europe and the British Isles.

In California, the first Europeans to see and write about the California Condor *(Gymnogyps californianus)* were Spanish missionaries, who called it variously the Royal Eagle or the Royal Vulture. It later became known as the California Vulture, a name that persisted into the twentieth century before being replaced by its present one.

Some of the more widespread raptors formerly were known by a variety of common names concurrently (at times depending on regional preferences), and sometimes the same common name was used for more than one species. Pigeon Hawk, for example, could refer to either the Sharp-shinned Hawk or the Merlin *(Falco columbarius),* both of which only rarely catch pigeons—unless they were named so because it was believed they looked like pigeons, which they do not. As recently as 1951 the Peregrine Falcon in North America was most often called the Duck Hawk because of its purported dietary preference (although many actually prefer rather smaller birds). All along, it was of course known to people taking more than a passing interest in birds that this species was actually the same as is found in Great Britain and elsewhere, where it has always been known as the Peregrine Falcon.

In order to make nomenclature more universally comprehensible and uniform, American ornithologists have in recent decades endeavored to bring North American common bird names in line with those used by the rest of the world. Thus, what formerly was called the Marsh Hawk here has now become the Northern Harrier *(Circus cyaneus),* another species represented both here and in the Old World. The American Kestrel, formerly known as Sparrowhawk (another case of mistaken identity because it preys mainly on insects and mice and only sometimes on sparrows) is very obviously a kind of kestrel, a falcon that is represented by a large number of similar species in the Old World. Conforming to general usage in ornithological books, the common names of the raptors in this volume are capitalized.

For greatest clarity in bird names, we have to turn to the scientific names used by the professional ornithological community. Ideally, the scientific name, or binomial (usually derived from Latin or Old Greek), should tell something about the animal—minimally, its affinities to other similar birds, or its color, or length of beak, and so forth. A binomial comprises two parts, the genus and the species. The Swainson's Hawk *(Buteo swainsoni)* and Red-tailed Hawk share the same genus name, indicating that they are close relatives, their differences being indicated by their species names. At their best, binomials describe the bird at hand: *Buteogallus anthracinus* literally means the coal black hawk-chicken, a quite descriptive name for the Common Black Hawk. But they can be exceedingly prosaic: the Golden Eagle is *Aquila chrysaetos,* which means "eagle golden-eagle."

Sometimes local varieties of a species are sufficiently distinct that they are given subspecific status, and a third scientific name is added, making it a trinomial—genus, species, and subspecies. The Peregrine Falcon has a distinctive Arctic race (subspecies) known as *Falco peregrinus tundrius,* and other Peregrine Falcons belong to one subspecies or another, although the population east of the Great Plains, once extinct, today consists of reintroduced captive-bred birds of mixed races and their offspring. A variety of races can be expected in any species that is very widely distributed.

Higher classification categories are more inclusive and indicate wider familial relationships. Caracaras are placed with the falcons in the family Falconidae because of their anatomical and genetic similarities. All diurnal raptor families, with the exception of New World vultures, are assigned to the order Falconiformes, the "falcon-shaped" birds. Officially, the New World vultures have been included with the Ciconiiformes, the "stork-shaped" birds, a placement based on anatomical and behavioral similarities and the results of DNA hybridization. However, more recent work indicates that these vultures are no closer to storks than to some other groups (J. Cracraft, pers. comm. 2002), and additional anatomical studies, supported by a reanalysis of the DNA data, seem to place the New World vultures as a sister group to those of the hawks and falcons (Griffiths 1994). This problem of evolutionary affinity remains unresolved.

Ultimately, what matters most to the hawk-watcher is the species, and unfortunately this concept sometimes can be problematic. For example, the Harlan's Hawk was at one time considered a separate species. Today, it is considered one of several races of the Red-tailed Hawk *(Buteo jamaicencis harlani),* because it fits within the parameters of what constitutes a species: a group of organisms that share a common gene pool and that freely interbreed under natural conditions and produce fertile offspring. So, although Harlan's Hawks are exceptionally dark and lack a red tail, they nevertheless interbreed freely with "more typical" Red-tailed Hawks and produce all sorts of perfectly fertile intermediates. The White-tailed Kite *(Elanus leucurus),* on the other hand, was "lumped" with the very similar Black-shouldered Kite *(E. caeruleus)* of Eurasia and elsewhere some time ago; curiously, a moderately astute observer, watching both species in the wild, could readily pick out differences between the two (such as tail length and flight style), and more recent studies indicate that they are in fact different species, thereby restoring the original common and scientific names to our kite. The American Ornithologists' Union is the ultimate arbiter in assigning names for North American birds.

A Hawk's Body

Weight reduction and streamlining are the leading themes in the construction of a bird's body. The entire skeleton is extremely lightweight, with the long bones being hollow and strengthened with fine internal struts; and, as any eater of chickens knows, the biggest muscles are packed around the large and deeply keeled breastbone, the center of gravity in flight. These are the muscles that in birds power the wings and make up about 12 to 18 percent of the bird's weight. For aerodynamic reasons, the wings themselves contain only very small muscles that are used for altering the shape of the wing, for aid in keeping it folded, and for changing the position of the wing feathers.

A bird's neck and legs can be neatly tucked into and folded

against the body to retain the body's teardrop shape in flight. This shape is further enhanced by internal air sacs, numbering eight in most birds. These membranous bags smooth out the contour of the body beneath the skin and feathers, and some actually reach into the wing bones and the skull sinuses. They also act as reservoirs and bellows that feed air to the lungs during flight at a tidal volume four times greater than in mammals (which have no such sacs), over a one-way path that permits continuous gas exchange during both inhalation and exhalation, thereby supplying large amounts of oxygen for the bird's metabolic needs. In addition, air sacs provide a means for the elimination of the substantial heat generated by the friction of the muscles during flight. In some hawks, the sacs may house great numbers of roundworms, acquired from their food, which usually do not visibly affect the bird, although they may sometimes kill it.

A bird's rather small lungs are at least 10 times more efficient than those of mammals. Fixed in size, they lack the inflatable balloonlike alveoli of the lungs of a mammal; instead, fine tubes (air capillaries) lie parallel to the capillaries of the circulatory system, and air passes through them in the opposite direction from the blood flow (a countercurrent arrangement), thereby permitting much greater gas exchange while taking up much less space.

As with other birds, a hawk's internal organs are crammed into a compact, rigid airframe. Although the relatively long neck is highly flexible (raptors have 14 or more neck vertebrae; we have seven), the rest of the vertebral column is not, in contrast to that of other vertebrates; obviously, a body in which the posterior half can flop from side to side or up and down would not be very stable in midair. And so the vertebral bones of the back and of the pelvis are fused, except for a single one that connects these two regions. In falcons, extra bones accessory to the tailbone allow for the attachment of powerful muscles that manipulate the tail during braking and other maneuvers at very high speeds.

A diet of meat places no great demands on a digestive system, and that of raptors is generally simple and short. Because proteins are so easily digested, even in large chunks, there is no need to grind the food, and hence the gizzard does not have a thick muscular wall, as does that of a chicken, but is instead a simple sac. There is a crop, a second sac at the base of the neck where sub-

Fig. 6. Sharp-shinned Hawk with full crop.

stantial amounts of food can be stored and carried about, a sort of built-in lunch bag; as the stomach empties, the hawk moves more food down with jerky lateral contortions of the neck and by pushing down. Hawks drink only occasionally, most commonly in hot weather and just prior to bathing; their water needs are largely satisfied by their water-rich diet.

Like all birds and many other vertebrates, hawks have a single outlet for all systems whose products must leave the body—the digestive, urinary, and reproductive systems. They all terminate in the cloaca (Latin for sewer), a chamber that opens to the outside world by way of the vent. The ovaries (of which only one is active) and the internal testes shrink when not in use, thereby reducing body weight in flight, and eggs are laid one at a time at intervals.

Beak

Raptor beaks come in a variety of shapes and sizes. An eagle's beak is massive and powerful; a Golden Eagle easily takes apart a Beechey Ground Squirrel *(Spermophilus beecheyi)*, an animal that has an unbelievably tough hide. By contrast, the beaks of the majority of New World vultures appear weak and are relatively slender, hooked probes that can be inserted into small spaces of large carcasses. Oddly, the largely carrion-eating Crested Caracara *(Caracara cheriway)* has a heavy beak, rather like that of an eagle.

All falcons have a short but thick beak capable of delivering a powerful bite. Falcons have a special projection on the upper mandible just back of the tip, the tomial tooth, which can be slipped between the neck vertebrae of the prey to snip the spinal cord, bringing instant death; these birds have much smaller rear and inner talons than other raptors of comparable size.

The White-tailed Kite has a small beak but an enormous mouth. The gape is so large that the hawk can swallow an entire vole or large mouse with ease, perhaps an adaptation for bolting down prey as quickly as possible; this slow-flying species is frequently robbed of food by Prairie Falcons (which may actually, on occasion, eat the kite itself). Another hawk with a great gape is

Fig. 7. Beak types: tomial tooth of a Prairie Falcon; small beak, large gape of a White-tailed Kite; elongate probe of a Turkey Vulture.

the Ferruginous Hawk *(Buteo regalis)*, a large, open-country buteo, in which this feature is thought to facilitate cooling off by panting, particularly as a young bird in the nest, which is typically exposed to direct sun. The large gape of this species is bordered by highly visible yellow lips, which serve as a readily seen field mark.

The beak of a raptor grows continually and must be worn down by bones, dirt, and other abrasive materials to keep it from overgrowing, which could ultimately lead to starvation. Captive hawks have been observed to bite into sand. A bare, easily cleaned patch of skin at the base of the bill, called the cere, surrounds the nostrils and is continuous with the lips; it is frequently bright yellow in adults and pale or even bluish gray in juveniles, which is also true for the skin of the legs and feet.

Most hawks have an oval nostril (external naris), but in the majority of falcons, it is roughly circular and has a central tubercle that makes the structure distinctive. There has been much debate about the function of this tubercle; for example, it has been thought that it serves as a baffle during high-speed dives, but some decidedly low-speed relatives of falcons also have this feature. In Ospreys, which dive into the water to catch fish, the slit-like nostrils can be closed. In Turkey Vultures, the nostrils are perforate; there is no septum, and you can see in one nostril and out the other.

Legs and Feet

The flexor tendons that clench a bird's foot run from under the toes up the backs of the leg bones, so that when a bird squats, the tendons must travel a longer distance, thereby automatically pulling the toes inward. At the upper end, these tendons originate from thigh muscles, which, when contracted, also pull the toes inward. In addition, the ends of the tendons under the toes have grooves that can interlock, ratchet fashion, with grooves in the sheaths that envelop them; a raptor's foot can thereby remain locked in a clenched position without outlay of energy. The downside of all this grabbing versatility is that a raptor may be unable to disengage its feet from oversized aquatic prey and consequently drown, a not uncommon fate of Ospreys.

The legs and feet of raptors provide an excellent clue to their diet, and they come in a variety of models tailored to their use.

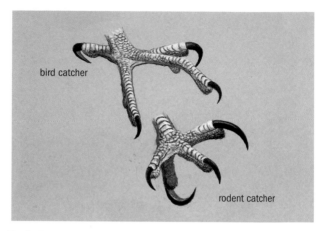

bird catcher

rodent catcher

Fig. 8. Foot types of raptors.

Hawks that predominantly feed on rodents or insects usually have short, strong legs and short, stout toes. They have a powerful grip, and as in most raptors, the talons of the rear and inner toes are conspicuously larger (enormous in eagles) and are used as daggers in dispatching prey. The other talons of the foot are smaller and serve to hold the quarry. Most raptors' talons are needle sharp. In Peregrine Falcons and perhaps also in some other raptors, the talons inflict not merely puncture wounds but small cuts as well, useful in expediting the victim's demise. Like the beak, the talons grow continually.

Bird-catching hawks have two kinds of legs but very similar feet. Aerial hunters' legs are short, whereas the legs of hawks that snag birds in bushes are slender and long. In both, the toes are long and thin, and some of the toes have soft, nipplelike pads on their undersides that evidently help to anchor the foot in feathers. The pads of a rodent catcher are flatter and hemispherical.

Vultures have fairly long legs and long front toes, but because the hind toe (hallux) is extremely short, vulture feet, while suitable for walking and scratching, are useless for grasping; a vulture cannot use its foot to catch or carry something. In addition, its talons, though sharp, are unspecialized and not very long.

Ospreys have long and strongly curved sharp talons of equal length, and the outer toe is reversible so that the whole foot can literally be wrapped around a fish; they have powerful legs for carrying sizable prey, and their toes bear tiny spikes to secure their slippery quarry.

Feathers

Feathers were long thought to be derived from the scales of the reptilian ancestors of birds. Recent work, however, convincingly demonstrates that feathers arose independently from tubelike skin outgrowths (Prum and Brush 2003). Numerous fossils of dinosaurs with feathers in various stages of evolution have been found, and the gradual elaborations of these structures are precisely echoed by the embryonic feather growth of modern birds. Feathered dinosaurs existed long before there were birds, and today's birds are now generally (though not universally) considered a group of theropod dinosaurs that developed powered flight.

Feathers are divided into distinct categories depending on their form and function. Like hair, they emerge from follicles and most, if not all, have small muscles attached to them and can be moved.

The overlapping contour feathers shape, protect, and insulate the body, and some (the long wing and tail feathers) also serve for flight; most can be raised and lowered, thereby varying the thickness of the heated air layer between the skin and the feathers. Specialized contour feathers known as coverts shape the airfoil of the wing; an upperwing covert arches up, whereas an underwing covert is flat. Whiplike filoplumes arise from follicles dense with tactile nerve endings and, being associated with flight feathers, may influence the movements of these. Fluffy down feathers provide added insulation. Semiplumes resemble both down and contour feathers: they have a long rachis, but their long barbs do not "zipper" together (see fig. 47, parts of a feather). Often found associated with the undertail coverts, they are sometimes used for display. Powderdown feathers crumble at their tips; they grow continually and are never shed. The powder they produce appears to aid in waterproofing feathers and causes the grayish bloom on some raptors' backs; a black velvet cloth wiped over the back of an adult Peregrine Falcon comes away gray. In some species, rictal bristles cover the nostrils and parts of the cere; these readily shed flakes of dried blood after feeding and keep dirt

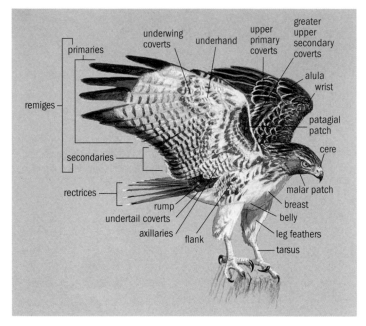

Fig. 9. Raptor topography (Red-tailed Hawk shown).

out of the nostrils, eyes, and so on. Gyrfalcons *(Falco rusticolus)* and Rough-legged Hawks, both Arctic raptors, have more feathers than hawks of comparable size from temperate zones. Ospreys have water repellent plumage and harder feathers on their breasts and legs, body areas that are first to strike the water in a dive, along with the beak and head.

As in other birds, the chief flight feathers of the wing (remiges) are divided into two groups, the primaries, which collectively arise from the outer, "hand" portion of the wing, and the secondaries, which are attached to the "arm." The secondaries are used chiefly to keep the bird airborne; the primaries are used for forward propulsion as well as for lift.

Whereas all birds' wings are modified to some degree, those of raptors are exceptionally finely tuned to serve special foraging methods. Raptors that frequently soar have outer primaries that

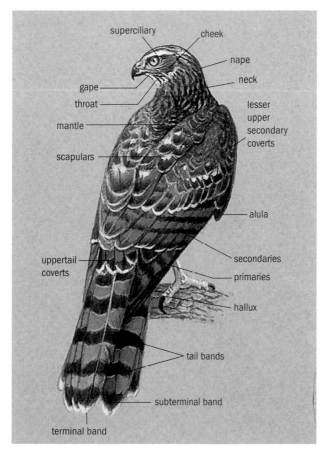

superciliary

cheek

nape

neck

gape

throat

lesser
upper
secondary
coverts

mantle

scapulars

alula

uppertail
coverts

secondaries

primaries

hallux

tail bands

subterminal band

terminal band

Fig. 10. Raptor topography (Northern Goshawk shown).

are deeply emarginated; that is, half or more of each primary is dramatically narrowed toward the tip (see fig. 46) so that when in flight overhead, the raptor appears to have widely spread "fingers" at the ends of its wings. These slotted remiges provide additional lift and are, in a sense, accessory "miniwings," or winglets, which, in effect, lengthen the wing. The added lift is derived from these

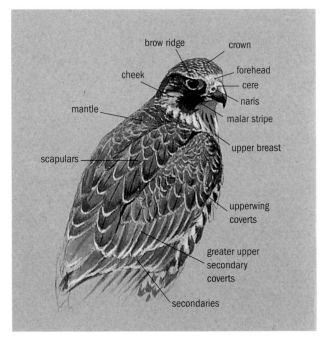

Fig. 11. Raptor topography (Peregrine Falcon shown).

winglets, reducing the drag caused by airflow toward the tips of the wings (Tucker 1993). Conversely, molting of some of the slotted primaries increases drag more than expected because of the resulting change of shape in the remaining slots (Tucker 1991). In raptors that pursue their prey by sprinting through dense vegetation, such as the Northern Goshawk *(Accipiter gentilis)*, the slotted outer primaries are extremely flexible and allow a deep and rapid wingbeat while providing powerful propulsion. Falcons, by contrast, which are less given to soaring, have long, narrow wings and minimal slotting mainly restricted to the outermost primaries.

A third group of somewhat less obvious flight feathers form the alula, attached to the "thumb" on the wing's leading edge. Alulae, when spread, furnish added lift during landing, and they can be extended during dives as stabilizers, like the fins of a rocket.

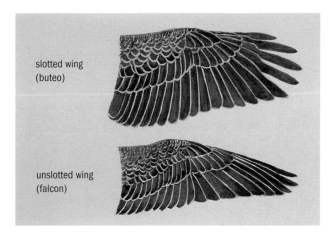

Fig. 12. Wing types of raptors.

Raptors normally have 12 tail feathers (rectrices) that can be spread widely or folded tightly. The extent to which a bird spreads its tail in flight is often correlated with the wingspread; in general, the more open the wings, the greater the tailspread, and with wings fully spread, the tail provides an estimated 10 percent of lift (Tucker 1992). The tail is also used for braking and for steering; a Golden Eagle, soaring in a wind, endlessly adjusts its tail's width and tilt. Various woodland hawks, not all closely related, have long tails to facilitate rapid changes in direction in tight quarters; highly aerial and soaring raptors' tails are often short because not much maneuverability is called for, and a long tail adds to drag.

Usually but not always, wing and tail feathers are longer in juvenile raptors than in adults; the reason for this difference is open to speculation. In very large raptors that do not molt all their remiges in one year, remaining first-year wing feathers can often be seen projecting beyond the newer, shorter ones in a second-year bird when the wings are spread. The central tail feathers of a captive juvenile male Peregrine Falcon, having been pulled out by a sibling, grew back with adult coloring and length, being nearly 3 cm (1.1 in.) shorter than the remaining rectrices.

The feathers of most raptors are pigmented, typically with shades of brown and black brown (melanins), and very often,

Fig. 13. Red-tailed Hawk attacking ground squirrels: the alulae are often deployed in fast glides, during stoops, and always during landing.

Fig. 14. Merlin with fully fanned tail to execute a tight turn as it mobs a Golden Eagle, whose tail is sharply angled as it maneuvers in a high wind.

Fig. 15. Differential damage to unpigmented parts of feathers. White bars damaged in secondary of Red-shouldered Hawk and rectrix of Prairie Falcon; white tip eroded from rectrix of male American Kestrel.

Red-shouldered Hawk   Prairie Falcon   American Kestrel

conspicuous bars adorn the flight feathers of the wings and the tail. The function of this barring is poorly understood, although dark pigment makes a feather stronger and more resistant to wear and, perhaps, to consumption by feather parasites. Thus, flight feathers are often tipped with black or dark brown.

Raptors that principally pursue birds in the air have fairly stiff flight feathers, whereas those of ground-game hunters are soft and pliable. This difference can be observed even in closely related species with quite similar foraging methods, such as Peregrine Falcons and Prairie Falcons, with the latter, as a partial ground hunter, having softer flight feathers. Likewise, the body plumage of aerial hunters is tighter and more compact compared to the fluffier covering of the ground hunters.

The plumages of juvenile (first-year) and adult birds are, with some exceptions, quite different, and some species also have subadult patterns (appearing after the first molt but before the full adult feathering is acquired, which in our eagles and condor takes several years). With Bald Eagles, second- and third-year birds look strikingly different from juveniles and adults. It has been suggested that juvenal and subadult plumages suppress aggression by older subadults and adult birds, while at the same time earn an older subadult more respect from younger birds.

Hygiene

After feeding, raptors clean their beaks by vigorously rubbing them, first one side and then the other, on the ground or on their perch, and vultures may similarly wipe their heads after reaching deep into rotting bodies. Some also clean their feet by wiping them against the substrate, at times crossing their extended toes while doing so; a female Golden Eagle was observed to pull her talons through her beak, then inspect the results of her efforts much like a woman would after filing her nails.

Bathing is good for feather maintenance and is especially important in birds that make their living by the chase; most raptors bathe regularly, some doing so daily, given the opportunity. Captive raptors may even bathe at night, with only starlight for illumination. Ambient temperature appears to be of no concern even though wet feathers that subsequently freeze can be fatal. Hawks can be extremely vulnerable to attack by bigger raptors while bathing, and some bathe only very briefly and furtively; soaked wings can render them flightless, especially the larger species.

Water softens feathers and restores their shape, and as the hawk dries after a bath, it preens: it runs the long flight feathers through its beak to help straighten them and adjust their webs; loose down is removed; and oil from the uropygial gland just above the tail is transferred with the beak to the various feather tracts and rubbed on. Hawks preen daily, not only after a bath, and often interrupt a preening session by "rousing," the raising and vigorous shaking of the entire plumage, or by stretching legs and wings, usually one side and then the other.

Some raptors, such as Prairie Falcons, also dust bathe, which may clean feathers as well as kill ectoparasites, the dust particles acting as abrasives between the plates of the arthropods' exoskeletons. All raptors appear to enjoy sunning themselves, not only for warmth but also probably for additional therapeutic effects; they usually turn their backs to the sun, fan their tails, and partially or completely spread their wings.

Fig. 16. Cooper's Hawk engaging in maintenance behavior.

Molt

Feathers become damaged and worn, and therefore all feathers must be replaced periodically, with the new ones pushing out the old. Molting is generally timed so that the absence of flight feathers does not interfere with migration or with the intensive hunting needed to raise young; however, built-in redundancy and various abilities compensate for such loss. Hawks can fly surprisingly well with damaged or even mostly missing tails. Broken and missing remiges, too, are usually not much of a handicap, although extensive damage here is more serious, particularly in the largest raptors.

New body feathers are usually in place by the time cold weather arrives. The Tundra form of the Peregrine Falcon begins to molt in summer in Alaska and northern Canada, interrupts the process during fall migration, and then resumes it on its Central and South American wintering grounds. Although both male and female raptors begin to molt their flight feathers during their breeding season at about the same time (at the onset of incubation), the female drops her flight feathers at much shorter intervals than the male, whose task it is to supply the female with food during much or all of the incubation period.

Fig. 17. Gaps in flight feathers caused by molting are often symmetrical, as in this Red-tailed Hawk.

In late spring and summer and into fall, it is very easy to spot molting activity—soaring raptors show conspicuous gaps in the flight feathers of both wings and tail. To grow a new flight feather takes about a month for a Redtail-sized raptor, less for a smaller hawk. In most birds, all feathers are replaced within a year, mostly during the summer, but in the largest raptors, the integrity of the wing is so critical that the wing molt is very slow and protracted and may last at least two years. In Golden Eagles, even some wing coverts may persist for three years before all are renewed, and the faded old feathers are easy to spot and facilitate the recognition of individuals. Raptors also molt the scales of their feet, much as they do their feathers.

Wings

The basic structure of a raptor's wing is the same as that of any other flying bird, that is, in cross section, the wing seen end-on forms an airfoil: the upper surface bulges near the leading edge of the wing and curves down toward the trailing edge, whereas the lower surface is flatter. As the bird moves forward through the air,

air molecules must pass over a longer distance crossing the upper surface than they must over the underside before reaching the wing's trailing edge simultaneously; the resulting pressure differential produces lift. Flapping increases this differential and, therefore, lift. Various other factors participate in the process that makes flight possible.

Three properties of the wing (camber, aspect ratio, and wing-loading) profoundly affect its function. The camber of the wing is the degree of the front-aft curvature; high-camber wings, with a conspicuous cross-sectional curvature, are very broad and suitable for prolonged, slow-speed soaring and provide much lift. Such wings are typical of vultures, eagles, and the various buteos (the latter are in fact commonly called broad-winged hawks). Low-camber wings, by contrast, are narrow and flat bottomed and enable the owner to travel at high speed while providing little lift at slow speeds. These wings are seen in falcons—and in jets. The shape of the airfoil creates drag at the wing's trailing edge, and birds can adjust the camber for various speeds to modify that drag; while landing, for example, large birds lift their upperwing coverts using small muscles, thus increasing the camber substantially to provide more lift and increased drag to slow down. Observant travelers can watch such changes in the camber of an airplane's wings during takeoff and landing.

The aspect ratio is the ratio of wing length (or wingspan) to wing width (the square of the wingspan divided by wing width). Wide wings, having a lower aspect ratio, provide much lift but also generate considerable drag; the long, narrow wings of falcons with their higher ratio cause less drag at the expense of reduced maneuverability.

The amount of weight that a unit area of wing must support (weight divided by wing area) determines what is called wing-loading. Raptors with a relatively low body weight and large, broad wings, such as Turkey Vultures, have low wing-loading. This is useful for prolonged, slow-speed soaring and is of course favorable to the conservation of energy. Heavy bodies supported by high-aspect-ratio wings (that is, a small surface area relative to the body weight they must keep aloft) indicate heavy wing-loading, suitable for high-speed pursuit in open spaces, as seen in many falcons; but these birds have to work harder at staying in the air. The forward thrust produced by flapping compensates for the additional drag.

Fig. 18. Cooper's Hawk flying upside down, with head right side up.

Fig. 19. Golden Eagle adult, about to land.

In active flight, the wingbeat of falcons is fast and snappy and often continuous, although these hawks also glide and soar. In a soar, a Peregrine Falcon looks like a toy bird kite; in a glide, its shape brings to mind a drawn bow. In a stoop, the wings are very nearly or completely folded so that the hawk looks like a missile. Thus, camber, aspect ratio, and wing-loading all can be changed to suit varying needs. Raptors can also fly at extremely slow speeds, pull in one wing at a time to pass between close trees, fly on their sides or upside-down, and rotate their wings into a vertical position as they " backpedal" to a soft landing on a branch.

Closely related raptors do not necessarily share all wing characteristics. The American Kestrel, a rather wimpy relative of the Peregrine Falcon, has much lower wing-loading; its less demanding prey does not require the large, powerful chest muscles needed for long-distance aerial chases. In the largest raptors, even extremely large wings do not always suffice: on windless days, California Condors, which are heavy bodied, have been seen to repeatedly hike up hills and launch themselves from the top in order to get enough air under their wings to stay airborne.

Flight

At takeoff, raptors jump into the air, run until airborne, or drop like a stone from their perch to gain speed. Humans have been thrilled since biblical times, and surely longer, by "the way of an eagle in the air."

Raptors in general are strong fliers; pragmatically, however, flight is mainly a matter of spending energy wisely. Gliding (point-to-point flight on set wings), the flight method most preferred by eagles and other raptors, is the most inexpensive, although no flight style is entirely free of energy costs.

A gliding hawk holds its wings out to the sides without flapping; the wings' surface area (and therefore lift) can be adjusted by partial folding. Because gliding usually results in a loss of altitude, a raptor wishing to stay aloft must find rising air that supplies sufficient lift to exceed the rate of drop. Updrafts along hill ranges and shorelines commonly support gliding without loss of altitude or provide a gain. The advantage of climbing in the air without a great outlay of energy also encourages raptors to make use of thermals, domed columns of warm air rising from heated ground, including parking lots or even steam-heated air over nuclear reactors; they may rise a mile high.

A thermal can lift a hawk, soaring (gliding in circles or ellipses), with truly amazing speed. Movements of the tail and primaries direct its course. The air in the center (or core) of a thermal rises much faster than along its edges, a phenomenon exploited by Sharp-shinned Hawks, which have a much smaller turning radius than large raptors. The biggest are restricted to areas where the air rises less rapidly, away from the core. The use of thermals allows a long, usually descending glide from the top while the hawk scans the land below for prey, winding up perhaps in another thermal. Great distances can be traveled in this fashion with a minimal expenditure of energy. There is some evidence that at least some soaring birds can "lock" their wings into place to further reduce energy demands. Thermals facilitate migration and also courtship flights in some hawks.

Powered flight, by contrast, is very expensive; in the American Kestrel, it requires about four times as much energy as does gliding (Gessaman 1980). Few raptors fly very far with constantly beating wings; buteos and eagles frequently do so when driving off territorial intruders, and falcons and accipiters do so in pursuit of prey, but even then they may interject some gliding. Most commonly, hawks alternately flap and glide in point-to-point flight.

In a vertical stoop (a headlong dive), the falcon may pump its wings to accelerate, or the wings may be completely folded against the body so that the bird is shaped like a teardrop; in Peregrine Falcons the body becomes extremely elongated, and sometimes the wrist of one wing projects further forward than the other, presumably to optimize the aerodynamic configuration. The air stream ripples the body feathers, the tips of wings and tail vibrating. Although eagles and some other raptors also perform these dives, the trim shape of falcons suggests speed, and indeed, it has recently been established that Peregrine Falcons, which are the most aerial hunters, can exceed 240 miles per hour in a stoop. This velocity was determined by releasing trained falcons equipped with speed-recording devices from airplanes at 12,000 ft and having them follow skydivers who then recorded their dives with high-speed cameras and camcorders. The falcons, thrown from the plane, would catch up with the latter and land on its struts, waiting for the trainer to jump. They would then pass their plunging trainer or trail behind to adjust their speeds to draw level with him. Peregrines clearly do as they please in the air.

Senses

The quality of a raptor's senses is also a reflection of its lifestyle. Hawks that forage in or over dense cover have excellent hearing, finely attuned to sounds indicating the presence of prey. The sense of smell is poorly developed except in the Turkey Vulture and some Neotropical raptors. The Collared Forest Falcon *(Micrastur semitorquatus)* of Central and South America, a bird of very dense vegetation, has been known to enter mammal traps baited with sardines, indicating that it uses olfaction while foraging (A. Reuter, pers. comm. 1998). Captive hawks demonstrate a sense of taste, sometimes turning down food that appears perfectly suitable, and they are often very quick to reject a piece of meat in which medication is hidden.

All raptors have excellent eyesight, and their visual acuity is legendary. Whereas humans have only a small central area on their retina for high-definition color vision, the cones (color receptors) of nearly all diurnal birds are spread all over the retina. For good measure, they have four types of cones where we have only three kinds. Many raptors, and perhaps all birds, can perceive ultraviolet. For example, the Common Kestrel *(Falco tinnunculus)* of Eurasia and the holarctic Rough-legged Hawk detect the ultraviolet reflections of rodent urine and feces and thereby locate vole-rich meadows (Koivula and Viitala 1999). The American Kestrel and the White-tailed Kite likely do so as well.

Contrary to popular belief, a hawk does not see as if looking through binoculars, but rather, owing to far more receptors, it sees objects in much greater detail than does the human. A mouse at 50 yards appears to a person as an indistinct, brown, oblong object; the hawk, from the same distance, sees the mouse as the same size but also readily distinguishes its eyes, ears, other details, and most important, movement. Each eye of the hawk has two foveas, especially acute focal spots where receptors are very numerous: the central fovea detects movement, while the temporal (side) fovea makes out detail. Captive raptors tethered outdoors can often be seen cocking their heads as they use one of their central foveas, tracking something (likely another raptor) across the sky, far beyond human visual range. Raptors are especially keen at picking out defects in potential prey, at times from a great distance—defects that make capture easier. This is done with the temporal foveas of both eyes, because hawks have binocular vision.

Fig. 20. Many a fine hawk photo has been marred by the nictitating membrane sweeping across the hawk's eye just at the moment of exposure. Juvenile Red-tailed Hawk shown here.

Like other birds, they also see faster than we do, that is, they can separate two visual images in about half the time that it takes a human eye to do so. Such "speed-seeing" enables some hawks to pursue birds in twisting flight through dense twiggery at an amazing pace. Many species bob their heads up and down or move them from side to side while looking at prey or potential enemies, presumably to gauge distance by triangulation.

Most diurnal birds of prey have a conspicuous brow ridge that gives them a fierce appearance; it may shade the eye from glare or protect it while diving into cover. As other birds, they have a nictitating membrane, a sort of semitransparent third eyelid that sweeps diagonally across the eye from the front and under the other two lids. It wipes the eye's surface and is probably also used in flight during rain and snow and perhaps in stoops.

Intelligence and Personality

A bird's brain lacks the mammalian brain's folds and wrinkles of the outer layer, the cortex, which we associate with intelligence; instead the brain's core, an area associated in birds with sensory perception and instinctive behavior, is highly developed and dominant. A falcon without cerebral hemispheres can grab a mouse but then is at a loss what to do with it (Welty 1975).

While much raptor behavior is stereotyped and instinctive, hawks are nevertheless clearly capable of rapid learning; they certainly learn very quickly when that process leads to a meal, which for a predator is nearly always an undertaking fraught with difficulties and sometimes danger unless the prey is grasshoppers or earthworms. The sport of falconry is based on this rapid learning ability. Raptors are always alert for clues that promise an easier meal—the limp of an injured rabbit, the distress call of a bird, or any physical irregularity in a potential prey bird, such as missing feathers. Oddly, the scream of a starling in the clutches of a Cooper's Hawk probably evolved not to put off the predator but rather to attract a second hawk that, while attempting to rob the first one, might permit the victim to escape. At many hunting clubs, hawks hurry to the sound of a gunshot and help themselves to a downed bird if it is small enough to carry off. A trained Cooper's Hawk, having once caught a pigeon in a barn, could not be carried within a hundred meters of the barn afterward without insisting on inspecting the building for another easily caught meal.

The practice of falconry has provided much opportunity to observe raptors' intelligence and personality, because captive hawks, and particularly those raised from chicks, soon become very tame and relaxed around their handlers (although they are often less so around strangers; they tend to recognize individual humans and dogs). If taken at a very young age, they imprint on the falconer and, regarding the man or woman as a mate, demonstrate almost the full range of their species' behavioral repertoire, although some of these behaviors may become exaggerated in the absence of proper socializing with their own species.

Different species and species groups have distinct "typical" personalities. The three *Accipiter* species tend to be high strung and quick to panic when faced with unfamiliar environmental stimuli; Red-tailed Hawks are placid and deliberate by contrast, as are Harris's Hawks *(Parabuteo unicinctus),* which, being very social animals in the wild, quickly realize the advantages of being with a human (and sometimes with a dog). Among Merlins, the Prairie race tends to be more cantankerous, and the Taiga form the calmest. Peregrine Falcons are calm but reserved, whereas Prairie Falcons seem friendlier but are given to outbursts of rage when offended; and Gyrfalcons are known for their playfulness and enjoy toying with tennis balls as they grow up and beyond. Amiable in their attitude toward their handler, Golden Eagles are appropriately unflappable and noble in demeanor, although they are known to give warning squeezes with their mighty feet. One eagle, using its beak, removed its owner's hat time and again and tossed it to the ground, perhaps in play or as a form of preening its "mate." Turkey Vultures, not exactly a falconer's favorite species, and California Condors can be almost doglike in their tameness and seem to become affectionate. Needless to say, such interpretations of raptor behavior are anthropocentric.

Wild raptors at times appear to show little common sense. A female Golden Eagle on her nest, visibly agitated by the approach of a human, quickly relaxed when the latter moved to a point equally close, where a large limb hid the eagle's head but not her body, so that she could no longer see the observer. A Cooper's Hawk several times bashed into a cage safely holding three pigeons, each time failing and going off to try from a different direction or using different approaches, such as frontal assault or sneak attack. The problem here was obviously that the hawk failed to understand that a wire cage is not a tangle of branches from which prey can be evicted.

On the other hand, the quest for food can reveal what appears to be intelligent behavior. An adult female goshawk, having killed a Blue Grouse *(Dendragapus obscurus)* in a mountain

Fig. 21. Gyrfalcon in a playful mood.

meadow, laboriously carried her quarry into a small stand of lodgepole pines *(Pinus contorta)*, where she began to pluck her prize. Suddenly she stopped, sleeked down, and, after staring intently at the nearby forest's edge, dragged the grouse under a pile of dead branches, then perched about 6 m (20 ft) away on a low branch. At this point, an adult female Redtail arrived and began looking for the grouse on foot for several minutes. Having failed to locate it, the Redtail flew off again, whereupon the goshawk extracted the grouse and continued plucking and ultimately fed on it.

With a few exceptions, most wild raptors are shy, although there is considerable individual variation; perched Red-tailed Hawks often allow a close enough approach, especially in a car, to take pictures with a good lens, whereas most Northern Harriers do not. Some Golden Eagles are amazingly fearless; other individuals flee at a quarter of a mile. Shyness in some raptors may be learned, at least in part. In the American west, the Northern Goshawk vigorously defends its nest, attacking humans that climb its nest tree or, sometimes, as much as walk near it. In Germany, where the same species was relentlessly shot until fully protected in the 1970s, goshawks were notoriously shy and never seen near their nests. With the cessation of shooting, this avoidance behavior rapidly disappeared, and today the goshawk is a common breeding bird in some German cities; the city of Cologne alone has about 30 nesting pairs, the hawks feeding chiefly on city pigeons and calmly ignoring passersby below their nests in city parks.

In California, much the same has taken place with the Cooper's Hawk, a formerly secretive species that, in central California, used to be confined to nesting chiefly in riparian and oak woodlands. In recent years, Cooper's Hawks have set up households in California cities; the city of Berkeley (Alameda County), for instance, is home to at least 12 Cooper's Hawk nests, some on branches overhanging streets, the young decorating parked and passing cars with their droppings. As ever more suburbs have spread into the margins of wildlands, Cooper's Hawks have raised their young often in plain view of humans, and these young in turn have gone on to breed in towns and cities. Here, too, the behavioral change is a result in part of the decline of hawk shooting and the presence of large prey bases. It also demonstrates the ability of certain raptors to adjust to new conditions and exploit new food sources, and to live more or less in harmony with humans in California.

Hawk Identification

Many raptors, especially those flying at a distance or seen only briefly, are often difficult to identify, and lively arguments sometimes follow a sighting by a group of bird-watchers, even of experts. It may be helpful to begin a discussion on identification with the numerous pitfalls that lead to bewilderment and misidentification.

Pitfalls of Identification

A lack of familiarity with raptor biology and the simple failure to note all of the field marks and flight traits are the most common causes of inaccurate raptor identification. Experienced hawk-watchers are amused by a novice's excited report of having spotted a white Gyrfalcon in oak savanna near Fresno, and carrying nesting material no less; that is because a beginner is unlikely to know that Gyrfalcons are exceedingly rare winter visitors, that no white one has ever been seen in this state, that they do not seek out oak savanna and do not nest here (ever), that they do not build their own nests, and that their bulk and flight style are very different from those of what the observer almost certainly saw: a White-tailed Kite.

A beginning hawk enthusiast observes a large brown hawk with a longish tail, flying by in a summer salt marsh, and decides it must be a juvenile goshawk. Juvenile Red-tailed Hawks, however, also have brown backs, and their tails are often considerably longer than those of adults and not red. But a Northern Goshawk has broad tail bands versus a Redtail's narrow ones; it has shorter, brownish primaries instead of long, dark-tipped ones, and a wide, conspicuous superciliary versus one that most often is inconspicuous. Add to that the habitat and time of year, and the goshawk identification is virtually certain to be erroneous (though there remains the very remote possibility that it is an escaped falconer's bird). In fact, the most likely candidate would be a juvenile or female Northern Harrier, provided it had the diagnostic white rump (see fig. 85).

However, experienced bird-watchers sometimes can be handicapped by their knowledge and fail to accurately identify a hawk because it is in an unexpected habitat or because they depend too much on color, an error which led expert birders to mistake dark morph Swainson's Hawks nesting in a Central Valley park for Golden Eagles. Obviously, size determination (which is often very difficult) was a problem, too.

Plumage variations in several species can be very confusing. The clear bib on the upper breast of a "typical" juvenile Red-tailed Hawk in California can be completely streaked in some forms, though the bib can still be made out. Some adult color morphs of the Red-tailed Hawk often look like they have been dipped into extra red brown pigment, but the basic pattern of the species can still be discerned under it all; the dark waistcoat and the much lighter breast are rarely completely hidden. Some light morph adults with all-pale undersides and dark morph adults, which may appear totally dark brown or black, force the observer to focus on field marks other than color or pattern, such as flight outline, flight style, and so on.

Unusual flight styles can cause confusion. Ferruginous Hawks clipping along with angled wings a few feet above the ground look almost exactly like very large falcons rather than the plump buteos that they are. Sharp-shinned Hawks and Merlins occasionally use undulating (sometimes called "bounding") flight in which the bird's flight path alternately dips and rises, the dips caused by the hawk completely folding its wings against the body before the next set of wingbeats. This is the standard flight of jays, most woodpeckers, finches, and other small birds and may represent aggressive mimicry by these small raptors. Turkey Vultures, famous for their dihedral wings, at certain times soar on almost perfectly flat wings, whereas their mimic, the Zone-tailed Hawk *(Buteo albonotatus),* which is very rare in this state, rocks like a Turkey Vulture in flight on dihedral wings and joins this species in cruising the countryside, again to approach unsuspecting prey. Turkey Vultures often flex their wings while gliding, but so do California Condors, Red-tailed Hawks, and Golden Eagles, albeit more rarely and the hawks and eagles much less deeply. Because of their exceptionally broad wings (for a falcon), Gyrfalcons can fly at a snail's pace, more slowly even, it seems, than a buteo in no hurry, but they can accelerate to speeds higher than those of any other raptor in the straightaway.

Because the flight feathers in most hawks become shorter after the first molt, proportions perhaps previously familiar to the observer can change. The wings may become more pointed, as they do in goshawks, lending the bird a falconlike outline in flight. Because of its diminutive size and rusty breast, a male adult Sharp-shinned Hawk, when seen from afar and in poor light, might be mistaken for a robin! But female Sharp-shinned Hawks are frequently confused with the similar male Cooper's Hawks (which they also resemble in size), although when actually measured, there is no overlap between the two species.

Common Nighthawk

Common Raven

Fig. 22. Nonraptors may be mistaken for raptors. The Common Raven *(Corvus corax)* is easily confused from afar with one of the dark raptors, but note the prominent beak and wedge-tipped tail; it is actually larger than a Red-tailed Hawk, although it does not appear so. On the Common Nighthawk *(Chordeiles minor),* slender, long wings with single, large white bars rule it out as a raptor.

We expect to see hawks perched in trees and on snags, telephone poles, fence posts, cliffs, and buildings, but most species at least sometimes sit on the ground (Prairie Falcons in the middle of a plowed field, Northern Harriers on levees, Ferruginous Hawks in a grassland, Golden Eagles on slopes), and so we fail to see them or dismiss them as inanimate objects or as species of less interest.

Whereas perched raptors are usually fairly easy to recognize as such, a variety of other flying birds are often mistaken for raptors. A lone gull (particularly a brown juvenile) or a pelican seen from afar can resemble a hawk, especially in a soar; Common Ravens *(Corvus corax)* soar also with set, flat wings and half-spread tail and appear quite hawklike except for the long, projecting head and beak. At a glance, even a pigeon or dove speeding by bears a resemblance to a small falcon. Nighthawks and swallows, too, can look like falcons at a distance. The usual hallmarks of most raptors normally clear up such errors—the large head, the tail length, the cadence of the wingbeat; all these help to set apart most birds of prey.

A report of a "kestrel" chasing shorebirds should be viewed with skepticism; "peeps" would be an extremely difficult quarry for a kestrel, but they are a very common prey of the similar-sized but much swifter Merlin. On the other hand, urban kestrels, especially males, do regularly catch songbirds, especially in winter, instead of the more traditional diet of insects and small rodents; they may also concentrate on songbird fledglings when they have young of their own to feed. Prey that is highly unusual for a raptor species should be cause to question an identification, but in itself should not nullify it; even the noble Peregrine Falcon has been observed to eat roadside carrion.

Hawks that have escaped from falconers could present real identification headaches to the hawk-watcher if such deserters were more common. Adherents of the sport sometimes use exotics, such as Lanner Falcons *(Falco biarmicus)*, Sakers *(F. cherrug)*, and Black Sparrowhawks *(Accipiter melanoleucus)*, all of them birds of the Old World. They also train captive-bred hybrids (see figs. 58, 102), which combine traits from both parent species, such as Peregrine and Gyrfalcon, Gyrfalcon and Merlin, or even Cooper's Hawk and Harris's Hawk (raptors belonging to different genera!), and they backcross such mixes or breed them with a third species, producing "tribrids". These "mongrels" have a very high hunting drive, which explains their popularity. Peregrine and Prairie Falcons in the wild have produced hybrids at least once (Oliphant 1991). Usually, hybrids greatly resemble one or the other of their parents or whichever species contributed the most genes; an occasional individual, however, can be truly baffling.

Field Marks

Seen through binoculars, that compact, apparently crow-sized hawk seen in profile atop the high-tension tower has a black-capped head and conspicuous broad, black malar stripes; its back is gray, and the underside shows black bars on a pinkish white background. The wingtips seem to reach the tip of the tail. Presently, the raptor launches itself in the direction of the shoreline, flying with deep, fast beats of the narrow, pointed wings; shorebirds, rising and joining together into a swirling cloud, signal its arrival in their midst.

This Peregrine Falcon has obligingly presented its telltale field marks: size, shape, color, and pattern; it then revealed its flight style, flight outline, and likely prey. Its perch and habitat, too, are typical for this species, although a Red-tailed Hawk, nearly the same size when perched, sometimes also sits on the same tower, making these clues less valuable.

Because not all of a raptor's field marks are equally useful, the dedicated hawk-watcher soon learns to look for all aspects of a raptor's persona, its "gestalt." The chief physical traits are obviously important, but so are flight style and the perches it uses, for example. Above all, the observer must keep an open mind and not jump to conclusions. It is embarrassing, having identified for fellow observers a Ferruginous Hawk by colors and size, to discover as soon as it takes wing that it is in fact a Redtail.

Size

Of all the observable traits, size is perhaps the most difficult to assess accurately. Distance and the absence of objects for comparison make size estimates highly unreliable. An enormous Golden Eagle soaring alone in the distant sky may seem no bigger than a Red-tailed Hawk. A hawk perched on a snag that projects above the crown of a leafy tree amidst the leaves often looks much bigger than it actually is, whereas a juvenile Red-tailed Hawk sitting in the top of a giant eucalyptus may strike the observer as no larger than a Merlin—especially when he or she really wants to see a Merlin. Also, various unrelated raptors are the same or nearly the same size, adding to the confusion.

Shape

Shape is very useful in sorting out a raptor's identity, whether the bird is perched or in flight. Most raptors can be assigned to one of three distinctive flight outlines; only a few do not fit easily into one of these categories. The silhouettes of raptors at rest also are sometimes very distinctive.

Raptors in Flight

"BUTEO" SHAPE: This shape is shared by vultures, most buteos, and eagles, all of which are frequently mistaken for eagles by the novice. The often-fanned tail may be short to moderately long; it always appears relatively short compared to the long, broad wings. In the Osprey, which also fits this pattern, the long wings are rather narrow. The buteo shape is perhaps the most familiar raptor shape.

buteo (Red-tailed Hawk)

accipiter (Cooper's Hawk)

Fig. 23. Outlines of
raptors in flight.

falcon (Peregrine)

"ACCIPITER" SHAPE: Raptors that hunt in the woods have long tails and short, rounded wings. Although wings and tail may be spread while soaring (which most woodland hawks also do at times), these traits are still prominent. Besides the three accipiters (Sharp-shinned Hawk, Cooper's Hawk, and Northern Goshawk), the Red-shouldered Hawk *(Buteo lineatus)* also has this shape.

"FALCON" SHAPE: Long, pointed, usually narrow wings and long tails are characteristic of the falcons and the kites. Other kinds of raptors approximate this shape when they angle their wings sharply backward.

A hawk can dramatically alter its wings' shape for various purposes. The long, broad wings of a Red-tailed Hawk, when nearly folded in a stoop, resemble those of a falcon, whereas the falcon's pointed wings become distinctly rounder at the tips when it soars. Also, recently fledged falcons may appear somewhat round winged because the longest primaries that give the wings their distinctive pointed shape have not completely grown in. The loss of several primaries during the midsummer molt lends an oddly pointed appearance to the wings of a Cooper's Hawk.

MISFITS: The Northern Harrier combines a long tail with long blunt wings and therefore does not neatly fit into any of these categories, and the rare Harris's Hawk, with its buteo wings and shape, has the long tail of an accipiter. In the even rarer Crested Caracara, all projecting body parts (neck, wings, and tail) are long, and the bird looks like a flying cross.

Raptors at Rest

Eagles not only are very large but also have conspicuously tall bodies from feet to shoulders when they sit upright. Perched with body horizontal, their legs appear too far forward, not at about the halfway mark of the body as they are in buteos.

Vultures often draw their heads down while at rest and from a distance may actually look larger headed than in flight because of a collar of feathers (ruff) that projects upward.

Buteos commonly look fat waisted because of their ample flank feathers that flare outward. Their heads are large and rounded.

Accipiters seem leggy, small headed, and long tailed. They have the habit of shaking their tails upon alighting and at intervals while at rest. Red-shouldered Hawks do so, also, but less frequently.

Falcons draw their large heads down, making the shoulders (actually bent wrists) appear to be hitched up very high. The wings are held tightly against the body, and the body appears less broad waisted than a buteo's. Falcons sometimes fold their tails to one feather's width, especially Prairie Falcons.

Windy conditions can alter the stance of any raptor as well as its outline. Facing into a high wind, a perched hawk's body is nearly horizontal. In wet or damp weather, many hawks partially open their wings, hanging them out to dry (wintering Red-tailed Hawks in the Central Valley show this behavior very frequently). In cool weather, the body feathers and those of the crown are puffed out for increased insulation, whereas heat causes the bird to flatten its feathers against the body. Overheated hawks are extremely sleek and pant rapidly. Young raptors often lie down to rest, not only in the nest but also elsewhere sometimes for a period after fledging, until the leg muscles fully develop.

Fig. 24. Back views comparing shape.

Red-tailed Hawk

Northern Goshawk

Peregrine Falcon

Color and Pattern

Color would seem a reliable trait, but in poor light, it can be very difficult to determine. Moreover, several hawk species are polymorphic, that is, they come in a variety of colors, and, colorwise, may look nothing like the usual form. Colors also fade; in the course of a year, chocolate may be sun bleached to pale tan. Hawks sometimes forage in burned areas and pick up enough soot to make them appear entirely brown black. White, which is not really a color, is a useful field mark in a few raptors such as the White-tailed Kite and the Osprey. Both, however, are commonly mistaken for gulls.

If the sun is behind the observer, colors are bright and helpful; if behind the raptor, they are useless and appear black. Patterns, by comparison, reveal themselves over much greater distances than do colors, even in poor light, and they tend to be far more consistent, even if the actual colors stray from the norm. The dark patagium surrounded by lighter coverts on the underwing of a soaring buteo, for example, at once identifies the bird as a

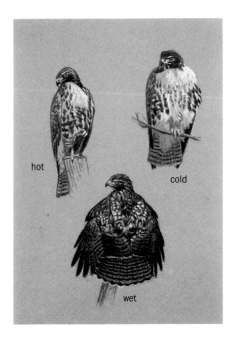

hot

cold

wet

Fig. 25. Red-tailed Hawk.

Red-tailed Hawk, whereas the basal white half of the tail, sharply contrasting with the dark lower half, quickly tells the viewer that the very large raptor overhead is a juvenile Golden Eagle.

Voice

A few hawk species not only have very distinctive calls but are also highly vocal, particularly Red-shouldered Hawks during the nesting season; such calls can provide instant identifications, but the hawk-watcher also needs a good ear: Steller's Jays *(Cyanocitta stelleri)* habitually mimic the most common vocalizations of both Red-tailed and Red-shouldered Hawks, often quite convincingly. The rough alarm calls of nesting Prairie and Peregrine Falcons can frequently be heard at their nest cliffs before the birds themselves are seen. Golden Eagles are a rather quiet species, but during nesting season, adults call at sundown to proclaim their territory.

Fig. 26. Red-tailed Hawks are not always easy to recognize. This unusually pale juvenile Red-tailed Hawk might be mistaken for some other buteo were it not for the dark patagial patch.

Flight Style

A hawk's way of flying is an important clue to its identification. A long-tailed, short-winged hawk that glides between bursts of quick wingbeats is either one of the accipiters or a Red-shouldered Hawk. A few leisurely beats followed by gliding low over vegetation indicate a Northern Harrier. Coasting on long, pointed wings that curve down a bit is typical of a falcon; at speed, it moves its wings with a deep, rowing motion. A similarly shaped White-tailed Kite flies much more buoyantly with softer wingbeats, unless it is headed into a stiff wind. Unlike the Peregrine Falcon, it is also given to frequent hovering or kiting. When Turkey Vultures beat their wings, the process often looks labored, and the body is frequently moved up and down. Eagles, on the other hand, have ponderous rowing, not laborious wingbeats, and their heavy bodies show no up-and-down motion.

Raptors soaring on a thermal can tell us a lot, too. Are the circles small or large? A Sharp-shinned Hawk usually makes much smaller ones than the similar Cooper's Hawk and is more likely to be tossed about than its more steadily riding cousin.

Interactions with other raptors also can provide clues; small hawks, diving at larger ones perched in trees or flying, are most commonly, but not always, American Kestrels, and during the nesting season, Red-tailed Hawks stoop spectacularly at Golden Eagles that enter their territory.

Perches

Raptors select perches either for temporary stopping points while foraging or for longer rests; a few use the same perch for both foraging and resting. However, it soon becomes obvious that many hawks prefer certain rest perches to others; Red-tailed Hawks are confirmed pole- or tree-sitters, and Rough-legged Hawks at times perch on extremely thin twigs. Accipiters like to conceal themselves inside tree crowns, although city Cooper's Hawks have an affinity for light standards. American Kestrels commonly sit on telephone lines, shunned generally by similar-sized Merlins, which usually sit either on treetops or in tree crowns but may also select other perches, such as fence posts. Bald Eagles like snags, whereas Golden Eagles relax on thick limbs in the crowns of big oaks. Large falcons can be found on high-tension towers (which are attractive to nearly all open-country raptors) and telephone poles, but also on cliffs and on the ground, and Turkey Vultures sit on snags and in eucalyptus trees. Although Swainson's and Ferruginous Hawks perch on poles, they are very comfortable on the ground as well.

AS PREDATORS, RAPTORS occupy the top position of energy pyramids in ecosystems; most have great mobility and require large foraging areas. Like other predators, they divide their days' activities chiefly between foraging, eating, and resting. A hawk with a full crop has little reason to fly about once it has found a safe place to sit, unless discovered by jays or other small birds made unhappy by its presence.

Conventional wisdom had it that the numbers of predators controlled the numbers of prey and vice versa. Although this balance concept has become controversial and is in fact sometimes incorrect, buildups of Beechey Ground Squirrels *(Spermophilus beecheyi)* quickly attract large numbers of eagles, and after a few weeks, the number of the rodents noticeably declines. Therefore, it appears that there can be a short-term effect. Prey density can also influence the breeding numbers of a raptor; the male Northern Harrier *(Circus cyaneus)*, for example, practices polygamy (polygyny) more often in years when vole populations are very high (although the females actually fledge fewer young than average). The vole population, however, fluctuates because of the food supply and other factors, not because of predation by harriers (Hamerstrom et al. 1985). Likewise, the White-tailed Kite *(Elanus leucurus)* is highly opportunistic and breeds in numbers during years of high populations of meadow voles, which suddenly appear in areas where these cyclic rodents have become numerous. When the Eurasian Sparrowhawk *(Accipiter nisus)*, which is a somewhat larger version of our Sharp-shinned Hawk *(A. striatus)*, sharply decreased in number in Great Britain because of the use of the pesticide DDT, there was no obvious consequent increase of songbirds, the prey of this species. Neither did songbird populations decrease when the Sparrowhawk population recovered after DDT use was discontinued (Newton 1986).

When raptors feed on highly mobile quarry, especially birds, their hunting success rate is remarkably low. It is the exceptional hawk that succeeds more often than not in the chase. All factors being equal, experience plays a major role. Adult hawks are far more adept at catching nimble prey than juveniles; some juveniles never get the hang of it and are quickly eliminated from the gene pool.

There is no doubt that, overall, predators tend to remove the old and the infirm from a prey population, along with the very young, the careless, and the stupid. Under the experimental conditions provided by falconry, it becomes apparent how readily in-experienced and uneducated prey is taken and how elegantly an experienced and clever veteran evades capture. However, because raptors are consummate opportunists, they sometimes succeed in taking healthy, experienced prey also, and anecdotal evidence suggests that some may in fact bypass weaker prey in favor of healthier though stronger quarry.

Hunting and Eating

Food Needs

Raptors are carnivores—even a kestrel that feeds on grasshop-pers is an eater of meat, a food that is not always easy to obtain. Compared with the vegetarian's diet of roots, stems, and leaves, meat has the advantage of being much more nutritious because it has more fats and proteins gram for gram, and, lacking a plant's hard cell walls, it is far more easily broken down. With very few exceptions, it is also free of the toxic or distasteful compounds that plants use for self-defense.

Desirable as meat is as a diet, it nevertheless has some decided disadvantages. Prey animals are many fewer in number and more thinly distributed than plants; they may run, hide, or even fight back. Meat spoils quickly—a raptor cannot simply eat one leg of a rabbit it has caught and come back a week later for another body part, whereas a deer might eat parts of the same rose bush in your garden week after week.

Obtaining a meal may require a considerable outlay of energy, and in some smaller raptors, the balance between the energy ex-pended in catching a prey item and the energy provided by it is very delicate. The two smaller accipiters, the Sharp-shinned Hawks and Cooper's Hawks *(Accipiter cooperii)*, are sprinters that usually pursue their quarry from ambush and quickly run out of steam. Juveniles especially may exhaust their energy re-sources during a prolonged storm (which limits their hunting) and perish. By contrast, raptors that leisurely soar while foraging (at times for hours) consume very little energy; for a healthy Golden Eagle *(Aquila chrysaetos)*, a week without food means lit-tle, especially in warm weather. In cold weather, a Sharp-shinned

Hawk needs to consume about 25 percent of its own body weight per day to fuel its energy and metabolic needs; an eagle requires only about 5 percent (Fox 1995).

Many raptors move with the seasons to areas of food concentration, in some cases actually following the food, as the Sharp-shinned Hawk does in the fall in the wake of migrating songbirds. Midsized raptors from harsh climatic regions, such as the Prairie Falcon *(Falco mexicanus)* and the Gyrfalcon *(F. rusticolus)*, can reduce their energy needs dramatically, presumably to survive periods when their environments make hunting impossible; they can "shut down" and wait out a snowstorm of several days.

Foraging Styles

No matter how low the daily requirement, much of a raptor's life is focused on securing food. A Golden Eagle often covers miles before all the various factors involved in a successful strike come together: a ground squirrel must be sufficiently far from its hole that, even though it sees the approaching eagle suddenly skim over a rise, it cannot reach safety in time—and it must not be agile enough to dodge the great feet. A sudden gust of wind must not spoil the final pivot as the bird throws back its wings and flings forward its feet, and there must be no intervening barbed wire fence or wind turbine that the raptor fails to recognize as something dangerous.

Foraging success in raptors is closely tied to the type of prey that they hunt and consequently to their hunting methods. Those that feed on relatively slow-moving or short-ranging prey such as grasshoppers and mice are sometimes called searchers; vultures, whose food usually does not move at all, are also searchers. Raptors that are called attackers feed on large, agile quarry; they have to overcome a set of physical problems not faced by searchers, and their foraging success rate is therefore much lower.

Prey vulnerability is as important as abundance; a mouse foraging under a blackberry tangle is safe from raptors. The seasonality of certain prey and the weather also influence hunting success. Foraging success rates should not be confused with foraging efficiency; a large meal repays the energy expended in the multiple attempts needed to obtain it.

Nearly every prey-rich habitat is exploited by one raptor or another. Typically, the predator is physically and behaviorally adapted to its hunting area. Because of the height and density of the vegetation, a forest or woodland habitat presents obvious difficulties to a raptor pursuing prey; prey animals are rarely in exposed places where they are vulnerable, and they are quick to take cover behind foliage, limbs, and tree trunks. Potential quarry consists chiefly of birds and squirrels (the small rodents here being mainly nocturnal), both quick to react to danger. Hawks living in such habitats (such as the accipiters in California) have evolved rather short, broad, and rounded wings that allow passage through narrow gaps between trees and branches as well as long tails that can be fanned instantly to make very sharp turns. These raptors, commonly called short-winged hawks, are capable of extreme acceleration, but most often they do not chase their quarry great distances. These wing and tail configurations are apparently so useful that they also appear in other, completely unrelated active forest raptors (Peeters 1963a).

Unlike hawks of the forest, raptors of grasslands and marshes spend much time on the wing. Here, the prey, chiefly rodents and insects, is often abundant, and although not swift, it is well hidden and spread out over large areas. Numerous birds of prey that are not always closely related have evolved to take advantage of this bonanza, and they have developed a variety of shapes and flight styles to overcome the problem of high-energy use while foraging efficiently over great stretches of land.

The hunting method most often observed is soaring, facilitated by thermals. Soaring is practiced by a wide variety of open-country raptors, most obviously by vultures, eagles, and buteos.

Another common very energy-efficient foraging method is kiting (hanging in the wind with almost motionless wings and little or no forward movement), which is used by a wide variety of raptors, including eagles. Kiting enables a hawk to position itself over prey-holding cover in the absence of a perch and to periodically move from one likely spot to another. The whole body of a kiting raptor acts as an airfoil, the camber of which can be altered by raising or lowering the tail.

Related to kiting but much less energy efficient is hovering (using intermittent or steady wingbeats and usually a spread tail to maintain a fixed, midair position). The hawk may "parachute" at a relatively slow speed down upon the prey or move some distance and hover again over more promising ground. Hovering is

Fig. 27. Red-tailed Hawk kiting.

most typical of the American Kestrel *(Falco sparverius)*, the White-tailed Kite, the Rough-legged Hawk *(Buteo lagopus)*, and the Osprey *(Pandion haliaeetus)*, but the Red-tailed Hawk *(B. jamaicensis)* and Ferruginous Hawk *(B. regalis)* and even the Prairie Falcon may hover. As the wind picks up, these raptors switch to kiting. Parenthetically, true hovering consumes so much energy that only very small birds can use this flight method, and then only for brief periods. Hovering in raptors is actually forward flight facilitated by a headwind of variable strength, during which a hawk may have substantial airspeed while remaining in one place, at zero ground speed.

Contour-hunting consists of covering large areas close to the ground in hopes of surprising quarry on the other side of a rise or a weedy ditch. Golden Eagles are masters at this, cruising at speed with barely a wingbeat, and hawks as diverse as Merlins *(Falco columbarius)*, Prairie Falcons, and Sharp-shinned Hawks all contour-hunt.

A similar method, quartering, is slower, requiring low wing-loading as the raptor covers large areas, often flying close to the ground, with few wingbeats and much gliding. A Turkey Vulture

(Cathartes aura) quarters over vast areas of woodland and open ground to locate carrion by sight and smell. The Northern Harrier also hunts over open ground but finds its quarry by sight and sound.

Ambush-hunting can be very effective. A male Prairie Falcon stationed himself on a high-tension tower on a hilltop near a cattail marsh and for several evenings flung himself at passing flocks of starlings as they came in to roost.

Still-hunting, a related form of foraging, involves waiting for unpredictable prey from a usually elevated perch, occasionally moving to new sites. The Red-shouldered Hawk *(Buteo lineatus)* is a confirmed still-hunter. This method is also in the arsenal of the accipiters, which most often still-hunt from perches or alternate short search forays with frequent perching; nevertheless, some also excel as aerial hunters. Where possible, most raptors approaching agile prey make use of intervening screens such as bushes, cars, or houses. Some hawks seek out favorite hunting perches if such are frequently productive. For years, a pair of Golden Eagles stationed themselves on a high-tension tower on a hill overlooking a pond and availed themselves of the ducks flushed by the periodic passage of trucks using the adjacent dirt road.

Aerial pursuit may be short and direct, as preferred by the short-winged hawks, or it may rake across the sky as a falcon stoops at its prey. The Peregrine Falcon *(Falco peregrinus)* sometimes stoops from enormous heights and captures prey high up in the air; other falcons, such as the Prairie Falcon and the Merlin, pursue prey into cover, much as less-aerial hawks do.

Many hawks are quite capable of using a variety of flight styles. The Sharp-shinned Hawk, which is usually a contour- and ambush-hunter, sometimes stoops at prey from a great height like a falcon, and the Ferruginous Hawk, a typical soaring buteo, has a low-level, high-speed hunting flight that can cause the observer to mistake it for a large falcon.

Seizing and Killing the Prey

For most raptors, attacking prey from above is clearly advantageous, although accipiters are very skilled at climbing after a fleeing bird and grabbing it from below. Falcons may swoop upward to seize the prey, having initially stooped below its trajectory. Of course, raptors that feed on insects and other arthropods can simply walk about in the appropriate habitat and snatch the prey with feet or bill, as the Swainson's Hawk *(Buteo swainsoni)* does in times of grasshopper abundance and the Red-shouldered Hawk does while picking up earthworms after a rain.

Most attacking hawks rotate their upper bodies and wings backward as they reach the prey and thereby increase the velocity of the legs and feet, which are flung forward. Large falcons in a stoop, on the other hand, typically angle their legs below the body parallel to the breast, so that their feet face forward. The quarry is either seized or struck a massive blow with one or both feet, which are open at the point of impact, not closed into fists as is popularly supposed. A handful of feathers from the prey may whirl up after impact. The blow may be forceful enough to kill the victim outright in midair or literally tear off body parts, even such firmly attached ones as a wing. An Osprey diving at a fish assumes a similar position (see pl. 2).

Although falcons chiefly kill by severing the spinal cord, true hawks kill by "kneading"; they drive their longest talons into the prey and then convulsively clench one foot and then the other alternately in an effort to penetrate vital organs. If a hawk is lucky enough to overcome a truly big item, so large that its talons cannot penetrate the vitals, it may dispatch the quarry by eating into it, aiming for the heart; the Sharp-shinned Hawk, for instance, kills doves in this fashion. While dispatching large prey, a raptor commonly braces itself with its wings to keep itself upright while the feet are otherwise engaged.

A hawk pursuing aerial prey occasionally falls into water with its quarry or, in the heat of the chase, plunges in after. It swims to shore by rowing with its spread wings and, if successful, tows its victim behind in a clenched foot. Some may even manage to rise from the water's surface, as an Osprey does routinely.

Prey that fights back may injure a raptor. A large duck can deliver a forceful and well-aimed blow with the bend of its wing, and a heron under attack can stab with its beak. A rattlesnake attacked by a raptor resists being eaten by lethal means, of course, although usually the hawk's feathers provide protection against the bites. A Redtail intending to make a meal of a rattler may land near it and spread and shake its wings so that the tips vibrate, inviting the snake to strike harmlessly. It then deftly snatches the distracted reptile's head with its foot.

Feeding, Mantling, and Caching

A successful hawk very often faces the danger of being robbed of its hard-won prize by another hawk; even eagles are not immune to such piracy. A young Bald Eagle *(Haliaeetus leucocephalus),* for example, may bully a Golden Eagle into relinquishing its quarry, sometimes for days, doing very little hunting of its own; but sometimes the roles are reversed. Juvenile and subadult Golden Eagles commonly try to rob one another as well, and even attempt to rob adults, usually unsuccessfully. Raptors often snatch at the prey another is holding or carrying without harming the owner; a juvenile Prairie Falcon, in a ground-skimming attack, robbed a female American Kestrel of a meadow vole while making no effort to seize the smaller hawk, which was perched with its prey in the grass. Sometimes two birds may grab each other's talons even in the absence of prey, a behavior called crabbing.

Falconers are keenly aware of how much robbery occurs in the hawk world. Anything attached to a tame hawk, such as leather jesses or a radio transmitter, often is mistaken by wild hawks as a prey item worth stealing. A trained male Merlin attracted the attention of a juvenile female Prairie Falcon, perhaps because at first she mistook the hawk's telemetry, attached to its

Fig. 28. A Golden Eagle (left) and its temporary kleptoparasite, a juvenile Bald Eagle.

leg, for prey he was carrying. Surprisingly, the Merlin continued his sparrow chases while the Prairie's attacks increased in intensity. At last, the worried falconer called down his hawk. No sooner had the diminutive raptor landed than its owner heard the sound of rushing air overhead. Looking up, he discovered an adult male Peregrine pulling out of a stoop that, without the man's presence, would likely have ended with his companion in the feet of the much larger falcon.

The size of the quarry influences where and how a hawk feeds. Small prey, such as grasshoppers, are usually swallowed whole, or nearly so, on the spot where caught. If the quarry is too large, the hawk has two options, both of them designed to prevent another predator from robbing the raptor of its meal. It can either take its quarry to an exposed perch where the approach of another hawk can be readily noticed and appropriate countermeasures taken, or it can hide with its prey in a furrow, in grass, or in some other cover, where it runs the risk of being discovered while in a very vulnerable position by not only other hawks but also mammalian predators. Hawks are, in fact, most vulnerable to other predators, especially other raptors, while distracted by wrestling with large prey; Red-tailed Hawks have killed many a Peregrine Falcon at such a time, and, conversely, have themselves been killed (and eaten) by Golden Eagles. Most raptors may eat other raptors, usually of lesser size. On the other hand, each year during fall and winter, a few falconry raptors are killed by larger wild birds of prey and not eaten, apparently to get rid of a food competitor; Golden Eagles are infamous for doing so, appearing especially intolerant of goshawks and large falcons.

Raptors themselves are acutely aware of the vulnerability arising from finding themselves below a larger, more powerful bird of prey. In winter, a Prairie Falcon and a Red-tailed Hawk may perch harmoniously atop a high-tension tower a few feet apart, and an adult female Prairie Falcon rested calmly a dozen feet above an equally calm adult female Golden Eagle. That same falcon, however, a day later, landed halfway up a pylon and shortly discovered a Red-tailed Hawk sitting high above her. Scolding loudly, she departed and alit atop the next tower down the line. On open ground, hawks the size of a Peregrine and smaller make themselves small and crouch over prey or even lie down on it, seeking to become one with the substrate when a larger raptor passes by.

Fig. 29. Red-tailed Hawk mantling. Note erect occipital crest.

At the approach or in the presence of another raptor of about equal size and especially of the same species, hawks "mantle" over their prey; that is, they spread their tail and wings over it and raise their back feathers and crests as well. Nestling hawks already display this behavior toward their siblings, as do captive-raised hawks, which mantle in the presence of their handler. Presumably, mantling not only hides the prey but also makes the defender look much larger. Bracing with open wings while dispatching large prey should not be confused with mantling.

Medium-sized and large hawks typically feed on the ground, a conspicuous exception being the Osprey, which appears to always use a perch of some sort. The short-winged hawks, if their prey is light enough in weight to move, seek the cover of a bush or other vegetation so as to feed unnoticed.

Eating is commonly preceded by plucking the quarry, although small mammals may be swallowed entire or in large pieces, skin, hair, bones, and all. Larger prey may sometimes be roughly skinned, as a Golden Eagle does with ground squirrels. Plucking, however, is most common, and some raptors, such as Merlins, are fanatic about removing nearly every feather before they eat. They, like all other raptors, readily form pellets, indigestible balls of fur, feathers, and so forth, which are later regurgitated (see fig. 45).

When food is easily caught or overabundant, some hawks may cache it, especially the various falcon species, but sometimes also the Red-tailed Hawk. Merlins may hide dead, whole sparrows in grass clumps by a rock, under shrubs, or among branches, both in their winter quarters and in their breeding habitat. Caching is especially common during the breeding season, when growing young require ample food; the male American Kestrel in some places visits colonies of Cliff or Bank Swallows *(Hirundo pyrrhonota, Riparia riparia)* to capture clumsy young on their first flight or by pulling them out of nest holes. After dispatching them, the kestrel then stows them near its own nest, to be eventually retrieved and delivered, as needed, to the mothering female kestrel.

It is not known for how long a raptor remembers a cached food item, but a captive Merlin, when released in the same area, made a beeline to the spot where it had hidden a dead sparrow six days earlier.

Resting

When resting, raptors seek out sites that are sunny or cool (depending on the season) and often those that offer some privacy or, in the case of the smaller species, protection from predators. Here, the hawk can attend to preening its plumage and even indulge in dozing with eyes closed.

Night perches are chosen with special care because of the dangers from Great Horned Owls *(Bubo virginianus)*. The Golden Eagle approaches its sleeping quarters leisurely and often perches nearby in the open for 15 minutes or more before retiring into the crown of the roost tree or onto a rocky ledge. Smaller raptors, however, use great stealth; a low-flying Prairie Falcon at dusk approached an array of wooden telephone poles at speed and shot up to a 10-cm (4-in.)-long steel foothold high up on one pole, immediately pressing its body against the pole, becoming one with it. It was not seen to move again as darkness fell. Another speedily flew into a narrow crevice in a cliff wall every night.

The Sharp-shinned Hawk, and likely the other accipiters as well, raises and spreads its scapulars when sleeping. Many feathers of these tracts (along with some inner secondaries and greater

Fig. 30. American Kestrel sleeping. The face is buried between and under the scapulars, not under the wing as commonly believed.

upper coverts) bear large white spots, which are partly or even fully hidden when the hawk is active but, when fully displayed, very effectively break up the bird's outline in low or dappled light.

Territoriality and Reproduction

Reproduction, the making of an individual that is the bearer of an organism's genes, is sometimes facilitated by territoriality, which tends to ensure that the process moves smoothly forward.

As with most other birds, territories are typically defended against members of a raptor's own species (intraspecific territoriality), although the owner of a territory may also attack and drive off individuals of other species if the latter use the same resources, such as food or nest sites (interspecific territoriality). A Peregrine Falcon may drive off a Prairie Falcon from its nesting cliff; one female Peregrine was seen killing a female Prairie in a stoop near a nest site (Walton 1978), but in another encounter, a Peregrine and a Prairie Falcon tumbled to the ground with locked talons and then separated, with the Peregrine fleeing the very aggressive Prairie (A. Fesnock, pers. comm. 2003). In a less-violent interaction, a resident female Prairie Falcon was observed physically pushing an intruding female Prairie off a ledge near her nest site, with the resident male then chasing the trespasser away (DiDonato 1992). Breeding Cooper's Hawks reacted most strongly to intruders of their own sex. Males most vigorously attacked trespassing other males, presumably to prevent possible mating with their females (which might result in the resident males having to raise the intruders' young). Resident females, possibly less concerned about the results of extramarital liaisons, chiefly drove off intruding females, perhaps to prevent them from usurping food supplied by the males (Boal 2001).

A trespasser of the same species is normally automatically the underdog and flees when attacked, although sometimes the intruder is sufficiently desperate to obtain a territory that it may kill the holder of one; this happens occasionally with Golden Eagles, for example, a species for which good territories with an adequate food base are often scarce.

Territoriality

Generally, defending an area for the purpose of raising young is of paramount importance; therefore, the first step in raising young is the establishment of a breeding territory. By definition, a bird's territory is an area defended for whatever purpose; it can cover several square miles or just a few feet in diameter, depending on the species. It may overlap completely or in part the raptor's home range, that area which supplies the raptor with food. When certain raptors nest semicolonially, as Ospreys do at times, the area holding and surrounding the nest may be no larger than a city block (the tiny town of Guerrero Negro in Baja California is home to over a dozen Osprey nests on telephone poles along its few streets), whereas foraging takes place over a much larger area some distance away.

Many raptors, however, have an all-purpose territory that not only holds at least one suitable nest site but also serves for courtship and copulation and supplies sufficient food for the raptor, its mate, and its young. Here, the territory coincides with the hawk's home range. The availability of sufficient food greatly influences the size of such a territory. In parts of Alameda County in central California, Beechey Ground Squirrels and suitable nest sites are so plentiful that Golden Eagles have territories that are exceptionally small for this huge raptor, measuring less than a square mile in some cases. This, in turn, has led to the greatest concentration of nesting Golden Eagles in the world here.

Raptors usually return to defend the same territories year after year, and some, such as Golden Eagles of the Coast Ranges, never leave, although territorial defense may stop entirely for a few months after the young have fledged.

A raptor drives away another raptor, or even a nonraptor, if it feels its eggs or young are in danger, though this is not territorial defense at all but rather harassing a potential predator. More appropriately, such activity is called mobbing, which is also commonly observed when nonraptors, such as songbirds and gulls, drive off passing hawks. Mobbing can be carried out by single birds or by groups and is a behavior used to drive off the perceived predator, especially (but not always) near the mobber's nest. Colonially nesting Red-winged Blackbirds (*Agelaius phoeniceus*), for example, conspicuously mob passing larger birds, such as crows, Great Blue Herons (*Ardea herodias*), and raptors, even Turkey Vultures and Golden Eagles, which are unlikely potential predators; they may even land on the back of the raptor and peck at it. A mobber typically stays above and behind the raptor; it is very unusual for even an agile hawk to turn the tables and succeed in catching its tormentor.

Most raptors exhibit some flexibility regarding suitable habitat for a breeding territory. The Northern Goshawk (*Accipiter gentilis*), which is generally thought of as a bird of dense coniferous forest, in fact nests in a variety of habitats, and the Peregrine Falcon lays its eggs not only on the traditional cliffs but also on many human-made structures in the state today, including busy bridges.

To proclaim their ownership of territories, raptors vocalize and perform aerial displays, the latter often serving simultaneously for drawing mates and for courtship (compromise behav-

Fig. 31. A Northern Mockingbird *(Mimus polyglottos)* mobbing a Red-tailed Hawk.

ior). Accipiters "sing" at dawn, as do Golden Eagles at sundown, uttering plaintive yelps that are repeated by neighbors in adjacent territories. Red-tailed Hawk pairs circling over their territories meet at the boundaries screaming what likely are obscenities at one another.

Normally, breeding territories are not defended year-round; however, some raptors such as the American Kestrel, having arrived at its winter quarters (which may be the same year after year), establishes a smaller winter territory to defend a food source. Generally speaking, hawks on their wintering grounds utilize a much greater range of habitats than they do for breeding; they go where the food is, and they are more accepting of other individuals, especially those of other species.

Sometimes such mutual tolerance leads to interesting encounters. A male adult Peregrine was seen stooping repeatedly, pendulum fashion, at an American Robin *(Turdus migratorius)* on the ground, the latter dodging and jumping out of the way much like a third-rate matador. Presently, a young Ferruginous Hawk joined in, making low, lumbering passes at the robin, and then a juvenile male Redtail, which was much more agile and could turn in tighter circles. The thoroughly unnerved robin now took to the wing and made for a fence some 200 yards away, with the Redtail in hot pursuit. The Peregrine followed, rose a few yards over the Redtail, stooped, and snatched the robin out of the air. This caused the Redtail to redouble its efforts, and, amazingly, it so pressed the falcon that it dropped its prey to the ground, where the Redtail instantly fell upon it and covered it, mantling.

Mate Selection and Seasonality

Just how hawks select a mate is not well understood, but most appear to mate for life (although they may separate when not breeding) and reinforce their pair bond each nesting season with courtship activities. Other arrangements, however, are possible. A male Peregrine Falcon may have more than one female, each with her own nest (as do some male Red-tailed Hawks). On at least two occasions, a Golden Eagle female had two males helping to feed her two young, an arrangement not uncommon in Harris's Hawk (Parabuteo unicinctus), a species in which two swinging couples may occupy a single nest or in which previous offspring may help in raising the current chicks (see species account). Juvenile nest helpers, presumably older offspring of the adults, have also been observed among Peregrine Falcons (Kurosawa and Kurosawa 2003). Monogamy notwithstanding, should a mate die or disappear during the nesting season, the surviving partner very quickly is joined by another, sometimes within a day or two, indicating the presence of numerous "floaters," individuals capable of reproducing that cannot find available territories. It is also likely that raptors are not so much faithful to their mates as to their territories, where, if they have migrated, the couples reunite for breeding (Harvey et al. 1979).

Although spring is generally the season for reproduction, the onset of breeding, signaled by courtship behavior, varies by altitude and by latitude in California as well as by species. Cooper's Hawks of the San Francisco Bay Area may have nearly fledged young while those of the High Sierra population are just hatching. Golden Eagles in central California may be on eggs as early as the first week in February, when Sharp-shinned Hawks in the same area have not yet contemplated courtship, or even sought out their territories. The principal reason for these variations in most cases is probably the need for each raptor to synchronize maximum food availability, generally in the form of young, clumsy prey animals, with the period of greatest need, namely, when it is raising its own young. In addition, the raptor's young need a supply of juvenile, inexperienced prey animals on which to hone their predatory skills.

For North American raptors, day length is normally the external cue to commence breeding activities, and that is the reason why hawks sometimes go through a false start-up of courtship displays in fall, when the day length is briefly equal to that of spring. On dreary days in late fall and early winter, resident Red-tailed Hawks are quite tolerant of juveniles in their territories and perch in close proximity. However, when the photoperiod equals that of early spring, a sunny day changes this tolerance into highly aggressive behavior. The weather appears to provide an end stimulus that promotes activities associated with breeding. An exception to this cueing may be the White-tailed Kite: young kites fresh out of the nest have been seen at all times of the year (J. Schmitt, pers. comm. 2002), suggesting that the photoperiod is of no consequence in stimulating breeding and that, for the kite, the important stimulus may be prey abundance.

Courtship

Distinctive and at times spectacular, the courtship flights of hawks are often noticed even by non-hawk-watchers. They may involve the male, the female, or both and often occur over or close to the future nest site. Among Golden Eagles, usually (but not always) the male rises to a great height, folds his wings, allows himself to drop at an angle or vertically for 50 m (150 ft) or so, then pulls out of the dive, and, after being carried upward by momentum, tips over into another dive upon reaching the stalling point, repeating the performance half a dozen times or more. Such aerial displays, various versions of which are found in most raptors, are called collectively "skydancing." Interestingly, because the same display is used to proclaim the territory, a female eagle may even do it from the nest, in much shortened form, upon seeing another eagle that is not her mate. As a courtship display, skydancing reinforces the pair bond and helps to bring the female into breeding condition.

Male Merlins and Peregrines roll from side to side in midair, alternately displaying their upper and undersides to announce their territories and attract mates. A male Cooper's Hawk per-

stoop   pull-up

Fig. 32. Golden Eagle skydancing positions.

forms shallow undulations, with very slow wingbeats and with the conspicuous white undertail coverts projecting from the sides of the tail base, and a male Turkey Vulture follows his chosen one closely in downward-curving swoops, tilting from side to side. Sometimes, male and female turn small circles high in the air and flutter their wings as they teeter. Male and, at times, female Northern Harriers fling themselves across the sky in extraordinary dives and loops, all along calling and flapping their half-folded wings.

Locking talons and spinning like a pinwheel toward the ground is part of the courtship of the Bald Eagle—although this display is also used by this species and by the Golden Eagle to discourage a trespasser from entering a territory. This behavior is not the same as crabbing, a locking of talons that occurs over possession of food or sometimes in apparent play. Male raptors may use courtship feeding, delivering prey to an appreciative female, which often responds with juvenile behavior patterns such as wing flapping and begging calls. The aerial prey transfer of the male Northern Harrier to the female during courtship can be spectacular.

In addition, pair members use other courtship displays, such as bowing and billing, as well as submission and appeasement displays. Copulation itself, the outcome of courtship displays, is very brief, a matter of seconds; it also reinforces the pair bond and occurs not only prior to egg laying but also, in most species, throughout the breeding season, even after the young are well grown. American Kestrels are truly heroic in this regard; pairs under observation in one study averaged 450 copulations each for the season (Villarroel and Bird 1994).

Nest Building, Egg Laying, and Incubation

California falcons and vultures do not build nests but use tree holes, hollow logs, crevices or nooks in rocks or buildings, ledges and pockets in cliffs or even mine shafts, the bases of boulders, and the "skirts" formed by the dead fronds of palm trees. All other raptors build their own nests, usually in sites distinctive for each species. The Red-tailed Hawk, for example, seems to prefer building its stick nest in large trees on slopes; but it may also settle for a big eucalyptus or valley oak *(Quercus lobata)* on flat land, or for a niche in a cliff face, where the nest, once used, may be taken over by Prairie Falcons the following year.

American Kestrels sometimes nest within a few dozen meters of one another, but most raptors insist on larger nesting territories. Both members of a pair contribute to building the nest, although the female does most of the work. Twigs and branches are typically torn or broken from trees rather than picked up from the ground, and their size reflects the size of the builder. Completed nests may be lined with bark flakes or with fresh greenery, which is renewed throughout the period of use of the nest, perhaps to discourage insects.

Raptors often build more than one nest. The Golden Eagle, for instance, builds (or repairs) several nests in its territory before it decides which one to use in a given year, and the White-tailed Kite makes a second nest in late spring (or perhaps even later) to raise a second brood. Some species, such as ospreys and eagles, may use the same nest and add to it year after year, resulting in huge, weighty structures that may eventually crash to the ground. However, the Golden Eagle, like most other nest-building hawks, may also build a rather small structure cunningly concealed in the dense crown of a tree. Only the Northern Harrier is a habitual ground nester.

Fig. 33. Swainson's Hawk carrying greenery to nest.

Nest-building skills vary with the individual; for years, a
Golden Eagle built or repaired low-quality tree nests that occasionally collapsed as the young grew. The eaglets survived the fall
and were raised on the ground. Distraught homeowners in Fremont (Alameda County), discovering that "their" Red-tailed
Hawk nest and young had fallen from their backyard pine, wired
up a cat basket a dozen or so feet up in the tree, where the hawks
promptly finished raising their brood (R. Britton, pers. comm.
2003).

Eggs are laid at two- or three-day intervals, depending on the
species. Clutch size (the number of eggs laid in a set) is a reflection of the food supply, competition, and survival rate, and
ranges from one, laid by the California Condor *(Gymnogyps californianus)* every other year, to as many as seven, the maximum
output of a few species, such as the American Kestrel. Long-lived
species that feed on scattered, large prey, such as the Golden
Eagle, generally have small clutches, laying but one or two eggs,
very rarely three; they may even skip some years entirely and not
nest at all. Smaller species with a very high mortality rate, commonly of the inexperienced juveniles, tend to have large clutches.

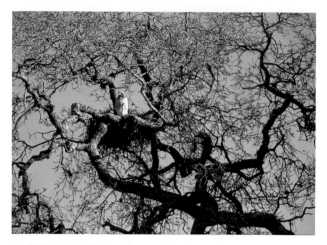

Fig. 34. Red-tailed Hawk at the nest. As a female hawk prepares to lay eggs, she often spends much time perched close to the nest.

Overall, the mortality rate is greatest in all raptors during the first year; if the hawk survives this, it is likely to live another four or more years (Johnsgard 1990). Because of the danger of high-speed collisions, raptors that are attackers live a life far more perilous than do searchers.

Some raptors, when the prey base is exceptionally high, nest very close together or engage in bigamy or polygamy. Still others, such as the Cooper's Hawk, appear to lay larger clutches in years of plentiful food. First-time breeders may lay fewer eggs than normal in their often poorly constructed nests. Most species "recycle" if the first clutch is destroyed by a predator or windstorm and lay a second or even a third clutch, generally in the same nest. Clutch size presumably compensates for mortality, the latter being greatly influenced by food supply and weather.

Incubation may start as soon as the first egg is laid, or more gradually as the clutch nears completion; falcons begin incubating after laying the next to last egg. Incubating raptors are extremely secretive, and even the Golden Eagle flattens itself into the nest to make itself less visible. The female bears the brunt of incubation duties; the male relieves her usually only

Fig. 35. Careful use of lethal feet around eggs and nestlings. Golden Eagle with recently hatched young.

for brief periods, and provides her with food. She also shelters the eggs and young from the elements and continues to work on the nest.

During the early stages of incubation, raptors are extremely sensitive to disturbances and may abandon a nest; as hatching approaches, abandonment becomes less likely, and a nest with young is rarely left. Some species, such as the Cooper's Hawk, mount a spirited defense, especially just after the young hatch and again as they approach fledging, and persons seeking to climb to the nest may be left with bloody scalps. Likewise, joggers enjoying their exercise in the forests of Lake Tahoe occasionally find themselves exposed to the wrath of heavily armed goshawks intent on defending a nearby nest.

Raptors are very careful with their feet when they settle on or turn their eggs, and to avoid breaking the eggs, they usually fold their toes forward, as they also do when brooding small young. The length of incubation is somewhat variable within a species and between species as well; it typically ranges from about a month to a month and a half, although the shorter spans are not necessarily seen in the smaller hawks.

Hatching and Fledging

After the young hatch, nearly all hunting is at first done by the male while the female guards the chicks. Some zealous males carry out this work with great enthusiasm; one Golden eaglet, the only occupant of the eyrie, lay on a bed of 52 dead Beechey Ground Squirrels, nearly all of them in various stages of decomposition.

The female first feeds the chicks minute pieces of meat, free of fur or feathers. As the young grow, they are given larger pieces and eventually are able to handle food that is only partly plucked. Vultures, however, regurgitate food for their young.

The urge to feed the offspring is so great that occasionally raptors raise, or at least temporarily feed, young that are not their own. A well-fed Redtail youngster was discovered in a nest also occupied by Bald Eagle chicks, likely the result of having been delivered to the nest as prey undamaged in capture and then having elicited parental behavior from the adult eagles (Stefanek et al. 1992). Even more unusual is the raising to fledging of a brood of European Starlings *(Sturnus vulgaris)* by a pair of American Kestrels that had discovered these aliens in a cavity they had hoped to use for raising their own family (Tlusty and Hamerstrom 1992). Clearly, certain traits possessed by chicks act as stimuli that bring out parenting across species lines.

Except for the largest species, young raptors grow fast. Dense white down replaces the sparser natal down and persists for the first two weeks; it is soon penetrated by the dark tips of the growing feathers of the juvenal plumage, most conspicuously by those of the wings and tail, which grow faster than the body feathers. At this early stage, the young are fed by the female, who attends to this task with great care and tenderness; the male delivers the prey to her, having first plucked it at a nearby "butcher block," the discovery of which is a sure sign of the presence of a nearby nest.

In the Golden Eagle and a few other species, one chick may kill another, an act known as siblicide or cainism. When eggs are laid several days apart, the first-laid will hatch before the other and produce a stronger and older chick that can overpower its younger, smaller sibling. Although it would seem that this is a mechanism by which all the food can be concentrated to produce a single, strong young, the Golden Eagle in fact often fledges two

Fig. 36. Down-tufted older nestling (Northern Goshawk).

or, occasionally, even three young that are as well fed and strong as an only child.

Once the young no longer require constant care, such as brooding to keep them warm, shading from intense sun, and protection from predators, the female can also go off hunting to help provide them with food. In raptors that feed principally on swift vertebrates (and especially in those that feed chiefly on birds, such as the Sharp-shinned Hawk and the Peregrine Falcon), it is the female that is the larger of the pair, a phenomenon that is known as reversed sexual dimorphism. It is thought that this size difference allows the pair members to take advantage of a greater range of prey, especially while raising young, with the bigger female feeding on larger prey, thereby widening the available prey base. The agile male, smaller by about a third in the more extreme cases, can on the other hand capture in greater numbers nimble, smaller prey, which is more abundant at a time when much food is required. It has also been suggested that the female's greater, intimidating size makes it easier for her to bully the male into surrendering its prey as he delivers it to the nest, and that she can fend off nest predators more easily. Large females, too, can theoretically commandeer efficient, small "trophy males" from smaller females. Reversed sexual dimorphism is also observed in raptors that feed on large, agile mammals. Overall, it appears that the less demanding the prey, the closer the sexes are in size. Among nestlings, the smaller males are faster to develop than their heftier sisters, presumably because they need to produce less bulk, and they typically are the first to leave the nest.

As the young hawks approach fledging, they exercise their wings with increasing frequency and jump up and down on the nest platform. At this stage, they are well feathered and have only a few tufts of down left, conspicuously on their crowns. The wing and tail feathers were the first to grow, but even with a head start, their wings and tails still look stubby compared with those of their parents. The youngsters are experienced at pulling apart prey delivered to them and at feeding themselves, and they watch with increasing interest anything that moves, from flies on the nest's edge to a deer walking by below. They snatch at branches and prey remains with their feet, honing their predatory skills. Soon the youngsters begin to venture out onto limbs near and around the nest: they have become branchers. Their bodies are fully grown, although they have still to gain some weight and complete the growth of their flight feathers.

The first flights of young hawks tend to be awkward and usually terminate in crash landings on lower branches or ledges or on the ground. Two fledgling American Kestrels, having accidentally dropped into a creek pool below their nest hole, quite capably paddled out like ducklings, using only their feet. Should a youngster succeed in laboring up to a higher perch, it may be afraid to come back down. Longer flights appear to terrify the young hawk, because it does not yet know how to land, and crashes are common but usually harmless. The adults patiently continue to deliver food and are noisily welcomed—hawks are never more vocal than when nearly grown or recently fledged, making certain that the food-providing adults can find them.

Many hawks do not hunt in the immediate vicinity of their nests, perhaps to leave vulnerable young prey for their own inexperienced young; the nests of Dark-eyed Juncos (*Junco hyemalis*), for example, have been found directly below those of Northern Goshawks. Some passerines such as House Sparrows (*Passer domesticus*) even build their nests in the sides of voluminous Osprey or Bald Eagle eyries, though they are of course unlikely prey for these large raptors.

Peregrine Falcons drop dead prey for their young to catch in midair, then live birds; eventually they will bring food but withhold it in order to encourage the young to move on. Golden Eagles, too, eventually refuse to surrender food to their complaining offspring.

Dispersal and Migration

Young hawks move away from the immediate area of their birth, a movement called dispersal. The young of some species tend to congregate. In mid- and late summer, for example, juvenile Redtailed Hawks sometimes form groups of five or six and together practice their flying and "footing" skills on breezy slopes, attacking clumps of grass, buzzing one another, and learning how to use the wind. Golden Eagle youngsters, having dispersed from the parental territory, associate, at times with subadults, in night roosts of a half dozen or more and relax together until warming air in midmorning makes it easier to start looking for breakfast. They, too, play together in the wind, much like young Redtails. Both species may toy with dried cow chips, which are carried aloft, dropped, and sometimes recaught in midair. Adults also sometimes play. High winds tend to excite Golden Eagles and cause them to engage in aerial displays, especially after a period of windless days.

Dispersal prevents eventual competition with the parents and is usually a permanent departure, although the dispersers of some species may keep coming back to once again beg for food, as do Golden Eagle young after having discovered how difficult it is to capture agile prey when you are very large and clumsy and inexperienced. Young that insist on returning to beg into late fall, when the following year's breeding season begins to get under way, may find themselves cuffed by their own mother in her efforts to drive them way.

In California, some adult raptors may also, after the young are gone, move away from the breeding area to other sites where food is more plentiful, normally over no great distances. Such a shift is called local movement, in contrast to migration, which is a regular movement away from and back to a breeding area, over distances of many hundreds or even many thousands of miles. Migrating Swainson's Hawks, for example, may fly from North America to Argentina, a round-trip of more than 22,000 km (14,000 mi) for some.

Birds that must use flapping flight to migrate travel mostly at night, when the atmosphere is calmer because of the absence of turbulence and thermals; however, migrating raptors (and other birds that soar, such as pelicans) ride thermals and so almost always travel during the day (Kerlinger 1989). Thermals rarely form over large bodies of water, forcing raptors to fly over land. Because of the funneling effect of the sharp narrowing of the continent south of Mexico, hawks migrating southward from all over North America become greatly concentrated there. Flocks of hundreds of thousands of some species are seen annually crossing the Panamanian isthmus.

In California, migrating hawks may follow the shoreline or take advantage of updrafts along hill and mountain ranges, which chiefly and conveniently run north to south. Because migratory pathways are so spread out in California and in the west in general, there are few conspicuous bottlenecks like the one at Cape May in New Jersey, which concentrates enormous numbers of migrating raptors. Migrants are particularly scattered in the north, where they must fly around multiple mountain ranges. But the Marin Headlands just north of the Golden Gate have a funneling effect, and a great variety of species in varying numbers pass through here with total numbers observed exceeding 36,000 in some years (A. Fish, pers. comm. 2003). On peak days, hundreds can be seen in an hour, and a good many individuals go unnoticed because they fly so high. Some migratory flyways are also found in the south of our state, for example in Kern County and along the San Gabriel and San Bernardino mountains (Los Angeles and San Bernardino counties). These latter sites have sometimes-large spring migrations, which elsewhere are more diffuse and not readily observed.

In the Sierra Nevada, the Northern Goshawk shows altitudinal movement, seeking out lower elevations, where prey is more abundant, for the winter months. The young of some species may leave the state, sometimes in unexpected directions. Young Redtails from southern California move northeast as far as Idaho and Utah and eventually return to nest near their place of birth (Bloom 1985), and some California-born Bald Eagles fly northward toward salmon concentrations (Hunt et al. 1992). For many raptors born outside the state, California is the final winter destination, and many species such as the Sharp-shinned Hawk and the Northern Harrier are consequently seen here in large numbers in the fall and winter months. For others, California is a transit point as long-distance migrants make their way from their northern breeding areas to their Latin American wintering grounds. Migrating hawks do not necessarily keep moving but may spend several days in one area if there is sufficient prey.

Raptors trapped and banded on the Marin Headlands in fall and then released are nearly as likely to move northward, and also eastward, as southward, and sometimes for considerable distances. One female Cooper's Hawk traveled about 400 km (250 mi) northward to the town of Mt. Shasta (Shasta County), and a juvenile Redtail was recaptured two and a half months later in Sonora, Mexico. On the other hand, a female adult Redtail was radio-tracked to a residential area in Marin County, where, it was learned, she, along with the young she raised yearly, had been fed store-bought chicken every day in a woman's yard for the past 18 years. For unknown reasons she sought out the headlands a few miles away in fall (Fish 1999).

How to Find Hawks

Whether on the floor of Death Valley or among the high-rises of San Francisco, hawks can be found nearly anywhere. They may lurk in the shadows of your backyard, circle overhead at your child's graduation, or perch on the road signs of a busy freeway. Often they are seen while searching for food; their visibility is deeply influenced by their food and hunting habits. At rest, some species sit completely exposed; White-tailed Kites *(Elanus leucurus)* shine from the tops of trees, and vultures sun themselves on snags and rocks. However, accipiters even at rest like to remain unseen for the most part and only rarely sit out in the open.

Hiking, biking, and even driving are all activities conducive to finding raptors. Some clues hardly bear mentioning; if a shadow moves across your path, it pays to look up. Many raptors have distinctive and sometimes stirring calls that beg to be investigated. The best hawk spotters, however, see more hawks because they know where, when, and how to look.

Where to Look for Hawks

Your home is a good place to start looking for hawks, especially if there is a songbird feeder in the yard (or in that of your neighbor). The nearly constant activity at one of these sparrow cafés acts as a draw for the smaller accipiters. Although lovers of songbirds generally do not like to see a finch carried off in the clutches of a Sharp-shinned Hawk *(Accipiter striatus)*, losses are always insignificant, and public attitudes have become more tolerant of such predation.

It is not uncommon to see hawks perched along city streets, whether residential, commercial, or industrial. In such settings, hawks sit not only on or in trees but also on streetlights and buildings. Light posts seem particularly attractive to Red-tailed Hawks *(Buteo jamaicensis),* Red-shouldered Hawks *(B. lineatus),* and Cooper's Hawks *(Accipiter cooperii),* whereas Merlins *(Falco columbarius),* American Kestrels *(F. sparverius),* and Sharp-shinned Hawks much prefer trees. Red-tailed Hawks sometimes also perch on buildings, along with Peregrine Falcons *(F. peregrinus),* which, however, usually require high-rises. All are attracted by the plentiful prey, often provided by cities, in the form of sparrows, starlings, robins, doves, pigeons, and rats. In some California cities, even Golden Eagles *(Aquila chrysaetos)* patrol the suburbs in search of squirrels and stray cats.

Hawk sightings in such areas commonly occur on your way to work, early in the morning, and that is prime time for raptor spotting elsewhere, too, because most hawks are foraging then. Some species such as American Kestrels and Northern Harriers *(Circus cyaneus)* are active throughout the day. Others such as the large falcons sometimes begin to hunt before sunrise; Peregrine Falcons have demonstrated that they readily feed at night on migrating songbirds in early fall (Wendt et al. 1991), and they are likely to do so at other times of the year as well, particularly during a full moon. Raptors that depend on thermals are more prominent between midmorning and early afternoon, and many raptors soar then as the day warms, especially when there is some breeze. High temperatures also encourage afternoon soaring as the birds seek out cooler heights. But noon is a slow time for hawk-watching; like other birds, hawks tend to be inactive in the middle of the day. Raptors also are less likely to be seen on overcast days, for lack of thermals. Cold and rainy days keep some prey species (such as squirrels) inactive, so hawks have fewer opportunities to hunt, and the rain may hamper flight. Generally, when prey is plentiful, most foraging hawks are seen in the morning, depending of course on the hawks' success.

Falconers have an advantage over bird-watchers who want to see hawks in the wild, because trained hawks in the air act like magnets that attract not only conspecifics but other species, too. Wild hawks, which normally would avoid a human, seem to appear from nowhere to drive away the avian interlopers from "their" hunting grounds. When not hunting themselves, captive hawks frequently gaze fixedly up in the sky, inviting the falconer to look up also, thereby revealing the passage of another raptor overhead.

Good places to find soaring hawks are hill ranges and bluffs, although heat rising from a large blacktopped parking lot can form fine thermals for soaring, too. Scenic overlooks and pullouts are frequently good places from which to spot soaring birds. The bluffs along much of California's coast and the mouths of northwestern rivers are excellent hawk-watching sites. Serious hawk-watchers scan the clouds with binoculars to find raptors flying so high that they appear as mere specks.

Fig. 37. Mouth of Pescadero Creek (San Mateo County). Tidal flooding promotes invertebrate abundance, which in turn attracts shorebirds, food for falcons.

In areas devoid of trees, such as parts of the Central Valley, the deserts, and the Great Basin, telephone poles often provide the only elevated raptor perches and are, of course, the places to look. Here, during the hotter part of the day, hawks find a sliver of shade as they press against the pole while also keeping an eye out for prey, which tends to be somewhat more abundant along roadsides, drawn by the denser vegetation here that has grown in response to rainwater runoff from the pavement. If telephone poles are lacking, raptors sit on fence posts, where they are often less conspicuous. Highways and roads are also popular raptor hangouts because of an often very diverse menu of roadkill—the dead and injured snakes, birds, ground squirrels, rabbits, and coyotes hit by speeding cars.

Raptors are more numerous at high elevations from midsummer to early fall as some hawks move upslope to take advantage of available prey and to nest before being driven down again by the first heavy snows.

Known migration routes can be very good places for finding raptors, provided the weather cooperates and it is the right time of year. The Klamath Basin, including the Tule Lake area, in northern California (November through March) and the alfalfa fields of the Antelope Valley in southern California (November through January) are especially good places to see and photograph wintering raptors. In the Sacramento Valley, the highest concentration of overwintering buteos can be found between Corning (Tehama County) and Stockton (San Joaquin County) (P. Detrich, pers. comm. 2003). The coast itself is productive, particularly north and some distance south of San Francisco. Fog is frequently a problem on the Marin Headlands, hampering both migrants and observers at this major hawk-watching site, where counts peak in late September but raptor activity is high year-round though most pronounced in fall.

Certain raptors, such as the Sharp-shinned Hawk, chiefly stick to north-south routes, whereas others are more scattershot as they leave their natal areas and strike out in many directions (Hoffman et al. 2002). Among these are Prairie Falcons *(Falco mexicanus)*, Cooper's and Red-tailed Hawks, and Bald Eagles *(Haliaeetus leucocephalus)* and Golden Eagles.

Migrants are greatly influenced by the weather, with some arriving just ahead of a cold front sweeping in from the north. Rising temperature and increasing barometric pressure following a weather front appear to promote movement, whereas headwinds result in much-reduced activity. Weather permitting, most movement appears to occur in late morning and early afternoon, confirming that hawks use warm air to their advantage (Hall et al. 1992).

Not all raptor species migrate at the same time, and there can be differences between juveniles and adults as well as between males and females. Migrating juvenile Sharp-shinned Hawks and Cooper's Hawks usually arrive in the San Francisco Bay Area in the last two weeks of September, but the adults come in mid-October and seem less dependent on the atmospheric benefits produced by the passage of a front. Juvenile Red-tailed Hawks appear in greatest numbers in early November, whereas the adults fluctuate little in numbers throughout the season, which runs from late August until the beginning of December (Hall et al. 1992).

The Central Valley and inner Coast Ranges serve as migratory pathways, and in late August to early September, flocks of Swainson's Hawks *(Buteo swainsoni)* numbering into the hundreds or even a thousand have been seen in Yolo County and in Kern

County at the southern end of the San Joaquin Valley and near the Grapevine. Kettles of up to a hundred Turkey Vultures *(Cathartes aura)* can sometimes be observed in fall and winter near Pleasanton (southern Alameda County), the birds gaining altitude on thermals before heading off northward or southward, depending on time of year. Thousands of these birds can be seen on peak migration days (usually around the end of September) at the Audubon-California Kern River Preserve at Weldon (Kern County) as they prepare to circle the southern end of the Sierra to head southward across the Mojave Desert. Many raptors pile up against the Transverse Ranges as they move southward down the Central Valley.

Of the various species, the Turkey Vulture is the first to drift northward from Mexico in the spring, with flights occurring as early as January. In general, the northward migration is much more spread out in time and space than the journey southward, and in some species, high winter mortality may reduce the number of birds returning. Usually, no great concentrations of raptors are seen, although a thousand northbound Turkey Vultures have been observed in the Imperial Valley (Imperial County) in late February and another flock of 2,000 in Lancaster (Los Angeles County) in early March. Concentrations of Turkey Vultures and Swainson's Hawks have also been seen near the San Bernardino and San Gabriel Mountains in March and April (H. Cogswell, pers. comm. 2002). Headwinds may cause lower spring coastal counts.

Near the end of April, most of the migrant raptors so obvious during the winter months have moved on, and by July and August, the species that breed in California have normally fledged their young; the adults are now deep in the molt, a period during which they, like all birds, are noticeably quiet and retiring. On the other hand, the young of some species, such as those of the Red-tailed Hawk, are much in evidence at that time as they practice their flying and hunting skills, often in the company of others. Subadults, floaters, and the young of Golden Eagles congregate more or less harmoniously at places holding high ground squirrel populations, because these sites are no longer being defended by resident adults. Young Cooper's Hawks undertake their first laughably clumsy raids at bird feeders. Generally, however, almost all raptors are much more visible in fall, winter, and spring, especially early spring.

Fig. 38. This dreary landscape of the southern San Joaquin Valley in winter is rich in rodents and rabbits and attracts many raptors during that season.

Finding Certain Species

Probably no one would have trouble finding a Turkey Vulture or Red-tailed Hawk in California, because these species are found in numbers the length of the state and in a variety of habitats, often gliding and soaring even over cities. Redtails are exceptionally fond of roadside perches, where they are very visible; in fact, buteos in general are famous pole-sitters, and, during migration, Highway 1 north of San Francisco seems to attract almost as many hawks as it has poles.

Some species, however, may be so rare or so secretive that finding them can become a major challenge. In no place in the state can a Northern Goshawk *(Accipiter gentilis)* be found with certainty unless a nest is known; you can hike for days in the High Sierra (the center of their breeding distribution) without seeing one. By contrast, Sharp-shinned and Cooper's Hawks are virtually guaranteed in the Marin Headlands and in the hills east of San Francisco Bay (for example, in western Contra Costa County's Tilden Park) at the right time (and right weather) in fall.

Certain hawks nest only in very specific regions of the state,

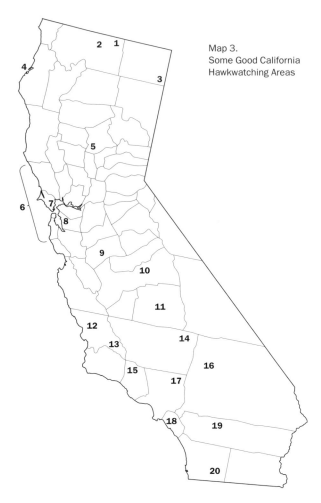

Map 3.
Some Good California
Hawkwatching Areas

1. Klamath Basin (Siskiyou and Modoc Counties): Hundreds of wintering Bald Eagles. Best chance of seeing a Gyrfalcon. Ospreys, Golden Eagles, many Northern Harriers, and Cooper's, Sharp-shinned, Red-tailed, Swainson's, Ferruginous, and Rough-legged Hawks.

2. Butte Valley (Siskiyou County): An agricultural area supporting both eagles, Ferruginous, Red-tailed, and Rough-legged Hawks in the winter and Swainson's Hawks in the summer. Meiss Lake Road southwest of Macdoel is productive.

Fig. 39. Northern Pacific coastal terrace (Salt Point shown).

3. Surprise Valley (Modoc County): Both eagles, Prairie Falcons, Ferruginous Hawks. Nesting Swainson's Hawks.

4. Arcata Marsh and Arcata Bottoms (Humboldt County): Redshoulders, Redtails, White-talled Kites, harriers, kestrels, Peregrines, Merlins, Bald Eagles, Ospreys.

5. Gray Lodge Wildlife Area (Butte County): Harriers, kites, kestrels, Turkey Vultures, Red-tailed Hawks. Wetlands of this and other areas of the Sacramento National Wildlife Refuge Complex also attract Bald Eagles, Ospreys, Ferruginous Hawks, Prairie Falcons, and Peregrines.

6. Northern and central Pacific coast: Coastal terraces and valleys of grassland and chaparral provide excellent raptor viewing, especially in winter. Red-shouldered and Red-tailed Hawks, Northern Harriers, kites, kestrels, Cooper's and Sharp-shinned Hawks, Ferruginous and Rough-legged Hawks, Ospreys, Peregrines. (Fig. 39)

7. Marin Headlands (Golden Gate National Recreation Area, Marin County): Exceptional numbers of migrating hawks. Most species counted each year. (Fig. 40)

8. Coyote Hills Regional Park (Alameda County): Northern Harriers, White-tailed Kites, Turkey Vultures, Merlins, Rough-legged Hawks (occasional), Redtails, Peregrines. (Fig. 41)

9. South San Joaquin Valley Wildlife Areas (Los Banos, Mendota, and Grasslands complex, Merced and Fresno Counties): Wintering Merlins, Red-tailed and Ferruginous Hawks, and Northern Harriers.

10. Millerton Lake (Fresno County): Wintering Bald Eagles.

11. Blue Ridge National Wildlife Refuge (Tulare County): All three accipiters can be seen in late summer (late August to early September) near this refuge in the foothills northeast of Porterville, although public access to the refuge itself is severely limited to

continued ➤

Map 3. *Continued*

Fig. 40. Marin Headlands. (Photo by Allen Fish.)

protect Condors (it is hoped that they will return to nest in this traditional summer roost area).

12. Morro Rock (San Luis Obispo County): Peregrine Falcons, especially in spring and summer.

13. Carrizo Plain (San Luis Obispo County): Northern Harriers, White-tailed Kites. A prime winter habitat for Golden and Bald Eagles, Peregrine and Prairie Falcons, and Ferruginous and Rough-legged Hawks. Pronghorns *(Antilocapra americana)* and Tule Elk *(Cervus elaphus)* have been reintroduced to encourage Condors to use this area. (Fig.42)

Fig. 41. Coyote Hills.

14. Kern River Preserve (Kern County): Turkey Vulture migration observation site.

15. Mt. Pinos (Ventura County): Accipiters. McGill Observation Point is a traditional Condor-watching spot.

16. Harper Dry Lake (San Bernardino County): Ferruginous Hawks and Bald Eagles in winter. Northern Harriers, Golden Eagles, Prairie Falcons. The lake is no longer dry.

17. Antelope Valley (Los Angeles and Kern Counties): Wintering Rough-legged and Ferruginous Hawks.

18. Upper Newport Bay Ecological Reserve (Orange County): Merlins, White-tailed Kites, Cooper's Hawks, Northern Harriers, Red-shouldered Hawks.

19. San Jacinto Wildlife Area (Riverside County): Golden and Bald Eagles, Peregrine and Prairie Falcons, Ferruginous Hawks.

20. Cuyamaca Rancho State Park (San Diego County): Golden and Bald Eagles, Ospreys, Northern Harriers, Turkey Vultures, and Red-shouldered, Red-tailed, Cooper's, and Sharp-shinned Hawks.

Fig. 42. Carrizo Plains.

but they can appear almost anywhere on migration and while dispersing. The Northern Goshawk, a bird that mostly nests in higher-elevation coniferous forests, is occasionally trapped nearly at sea level by banders on the Marin Headlands and is sometimes seen in the oak woodland of Alameda County and even in the salt marshes of San Francisco Bay. Obviously, there is a greater chance of seeing a Northern Goshawk in the Sierra in pine forest or aspen *(Populus tremuloides)* woodland above 1,800 m (6,000 ft), but like many other raptors, this hawk may turn up in places where it is completely unexpected, a reflection of the great mobility of most birds of prey. The Gyrfalcon *(Falco rusticolus),* although it does not nest here, may move temporarily into California because of food shortages to the north.

Announcements of unusual hawk sightings are often posted to online birding group message boards and are included in local Audubon Society chapter newsletters (some of which can be read online); an Internet search for specific uncommon raptors can yield exact recent sighting locations. Distribution maps of the more common raptors can be found on the Internet as well, although these are very general and, often, are works in progress. The North American Breeding Bird Survey results, for example, may be accessed at www.pwrc.usgs.gov/bbs.

Arnold Small (1994) provides a more thorough treatment of locales; unfortunately, some of the information provided by this excellent reference is inevitably dated, incomplete, or both, again reflecting the dynamic nature of bird distribution and the limited usefulness of supplying actual observation sites. Fluctuations in prey availability, construction of new housing developments, the season, and the weather are among the various factors that may temporarily or permanently change a site's reliability for holding a desired species.

Finding Raptor Nests

It can be exciting to find a hawk's nest, but it is very important to remember that it is against the law to disturb nesting raptors. Some species are easily frightened and will abandon their nests, although they are less likely to do so once the young have hatched. A few kinds of hawks are sufficiently tolerant to allow distant observers to watch the comings and goings of the parents as they feed the young, and that is particularly true for city nesters; others, however, such as most Golden Eagles, are exceedingly private about their home lives. Climbing trees to look into raptors' nests can also have dire consequences for its contents: Raccoons *(Procyon lotor)* habitually follow a human's scent to the nest and eat the eggs or young.

Often, the display flights of adult hawks terminate at or near the nest and thereby give away its location, as with White-tailed Kites and the accipiters. By contrast, American Kestrels, their vocalizations and aerial displays notwithstanding, are usually exceedingly secretive as they enter and leave their nest cavities. Red-tailed and Swainson's Hawks can often be seen carrying nesting material and, later, food, thereby giving away the nest site. Red-shouldered Hawks draw attention to their presence by their noisy vocalizations, and most raptors utter protest or alarm calls when humans get close to the nest; such vocalizations become most vigorous when the young are about to fledge. The begging calls of young, too, can frequently be heard over a considerable distance.

Some raptors place their nests in very conspicuous sites; the Osprey *(Pandion haliaeetus)* likes to build on snags in plain view, and a Bald Eagle nest that is 2 m (6 ft) across and 3 m (10 ft) deep is hard to miss.

The approximate location of the ground nest of a Northern Harrier is quickly given away by the comings and goings of the adults. Inspecting such a nest should be avoided at all costs, because not only do raccoons and foxes follow the human scent trail but also dogs, crows, and ravens, watching from afar, are inclined to find out just what was so interesting—and help themselves to a meal of eggs or young.

It can be exceedingly difficult to find the nests of the three accipiter species in their woodland or forest habitat without keeping a very sharp eye out for circumstantial evidence. These birds are very secretive, and even after the nest has been found, they may not show themselves at all, depending on how far egg incubation has advanced or how old the young are. Typically, each accipiter has its own set of criteria for choosing a nest site, although each is also quite flexible. Nevertheless, to find a goshawk nest for example, it makes sense to start in the right habitat and look first for the constellation of elements that goshawks find attractive.

Professional biologists sometimes walk the woods with tape

Fig. 43. On the eastern slopes of the Sierra Nevada, aspen "stringers" are typical nesting habitat for Northern Goshawks and Sharp-shinned Hawks.

players and recorded accipiter calls when "testing" a forest for the presence of a nest, and at certain stages of the nesting cycle, resident accipiters may respond to these auditory probes. Unfortunately, this method has caused some birds to abandon their nests, believing perhaps that the calls came from stronger interlopers.

The nests of tree squirrels are sometimes mistaken for those of Sharp-shinned or Cooper's Hawks, but the former are more spherical and always incorporate dead leaves in their walls; hawk nests are flatter and soon are decorated with down, visible from below, shed by the incubating female and later the young.

The begging calls of fledged juvenile Sharp-shinned Hawks sometimes serve biologists to locate their home nests in central California, and merely glimpsing such birds in late July and in August is good evidence that their parents nested nearby. By contrast, a juvenile Bald Eagle at that time of year may have fledged long before and moved hundreds of miles.

A hiker following a creek bed in the Coast Ranges may be startled by the sudden flapping of large black wings as a Turkey Vulture leaves its nest, which may be a mere depression against the base of a large rock. It is amazing how a Turkey Vulture, so conspicuous in the air, seems to vanish whenever it is near its nest.

Fig. 44. Prairie Falcon young, near fledging, in an eyrie (using an old stick nest of a Red-tailed Hawk), showing copious whitewash.

Scanning cliffs with a scope or binoculars often reveals either the cliff nest sites of large falcons, the birds themselves, or enough circumstantial evidence to suggest that a nest is nearby.

Circumstantial Evidence

Raptors frequently betray their presence indirectly and call attention to their presence before we see them. A careful observer can learn to keep an eye open for these clues.

Whitewash

White streaks of excrement ("whitewash") on cliff faces may indicate the presence of cliff-nesting raptors if it is spring, and copious amounts of whitewash splashed on the pavement of a street or sidewalk usually indicate the nest of a Red-tailed or a Cooper's Hawk overhead, or perhaps that of a Red-shouldered Hawk. At any time of the year, the presence of whitewash may simply mean that you have found a favorite roost. Dedicated hawk-watchers soon develop skills in identifying the different varieties of whitewash.

All birds evacuate digestive and urinary wastes simultaneously, which accounts for the two-toned appearance of bird droppings. The white part, coming from the kidneys, is chiefly composed of uric acid (white because it is a crystal, like table salt); it is the waste product of protein breakdown and deamination. The dark portion (usually black, brown, or green) is digestive waste from the bowel. Because birds of prey excrete some water with those wastes, the dropping of a hawk or owl often has the appearance of a fried egg. You can determine the general source of such a dropping with some confidence: if the dark part (i.e., the "yolk") is elongate and shaped a bit like a little black worm, the dropping came from a small owl; if the dark part is broken up and partly mixed with the white ("scrambled"), the originator was likely a large owl or a vulture. Vulture droppings, in addition, are often very watery, with the white portion not nearly as bright and flaky when dry as in the case of hawk droppings (or thick and crumbly as in those of owls). The droppings of hawks, eagles, and falcons, finally, usually have a well-defined, more or less circular dot or oval lump as the "yolk." A greenish color to this dot indicates that the raptor's digestive system was nearly empty.

The output of whitewash under an urban Turkey Vulture roost is spectacular but quite different from the almost equally impressive splash area under a Cooper's Hawk nest; in the vulture's fallout, the white portion is often fragmented into many white specks because of the great height from which the dropping was loosed.

If the whitewash is in the lower third or so of a cliff, it is likely from a Barn Owl *(Tyto alba)*; if in the upper quarter, it could be from a Prairie Falcon or a Peregrine, the latter often choosing cliffs of great height with a river at their base or other open water nearby. Barn Owls also may perch or nest occasionally in the upper parts of a cliff, however.

Whitewash not protected by overhangs usually is removed by rain, so in those parts of California with winter rain, new accumulations indicate an active site. In arid climates, whitewash persists on cliffs, oxidizing pinkish with age, making new nesting activity harder to spot.

Pellets

Inevitably, a feeding raptor swallows at least some of the feathers or hair of its quarry; at other times, the prey may be eaten entire, fur and all, especially if it is small compared to the hawk. All indigestible materials, including accidentally ingested grass blades, leaves, sand, and such, are compacted in the stomach into a pellet, or "casting," which is, depending on its size, typically regurgitated one day after the meal.

Pellets are usually oval, or elongate with rounded ends, and smooth. When fresh, they are covered with a thin, clear, elastic layer of mucus through which the compacted hairs or feathers can be seen. They are often found under roosts or favorite perches, such as high-tension towers, and a hawk's pellets are easily told apart from those of owls, which also produce them. Owl pellets are full of bones, sometimes including entire skulls and nearly whole skeletons, whereas those of diurnal raptors only occasionally contain fragments of the largest bones, a reflection of the more potent digestive juices of hawks. An exception is the pellet of the White-tailed Kite, which contains small bones and skulls, like that of an owl. Bone, incidentally, contains large amounts of usable proteins, reportedly, ounce for ounce, more than meat (which is full of water) and is therefore valuable food if the animal can digest it. Some prey remains are more conspicuous or persistent than others, so some species of prey are overreported in a pellet analysis.

Molted Feathers

During the molting season, an observant hiker can find raptor feathers in a variety of habitats, although they seem to most easily catch the eye in woodlands. Most large, brown feathers found away from water, especially when they are barred, belong to birds of prey, a notable exception being the stiffer and more curved wing feathers of the Wild Turkey *(Meleagris gallopavo)*. These discarded structures are proof positive of the presence of a certain species of raptor if the finder can identify the feather. It is highly unusual to find a molted feather of one of the very secretive accipiter species, the Sharp-shinned Hawk, Cooper's Hawk, and goshawk, unless there is a nest nearby, assuming the habitat is appropriate. Several molted feathers in one place usually indicate a favored perch or a night roost. Feathers lost during a struggle with prey on the ground tell a story when found days or even weeks later beside the remains of a victim.

A number of museums in the state display stuffed hawks, and here the hawk-watcher can become familiar with the plumages of the various species. It is illegal, however, to possess raptor feathers of any sort (or any other parts) without a valid permit or fed-

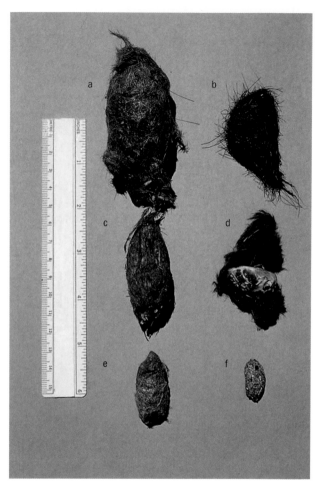

Fig. 45. A sampling of pellets: (a) Golden Eagle; (b) Turkey Vulture; (c) Red-tailed Hawk; (d) White-tailed Kite (opened to show the intact vole skull); (e) Peregrine Falcon; (f) American Kestrel.

Fig. 46. A sampling of wing feathers: (a) Golden Eagle right primary; (b) Turkey Vulture left primary (note that rachis is white its entire length); (c) Red-tailed Hawk left primary; (d) Northern Goshawk primary; (e) Prairie Falcon primary; (f) Peregrine underwing covert (note vanes of approximately equal widths); (g) Red-shouldered Hawk left alula feather (note very short calamus, typical for feathers from this group); (h) Red-tailed Hawk upper greater primary covert (note long calamus, typical for these feathers). Note that (a) through (d) are deeply emarginated, whereas (e) is barely emarginated; (d) through (h) are dorsal views.

Fig. 47. The parts of a feather, shown on a Red-shouldered Hawk tail feather (rectrix). The barbs are "zipped" together to form vanes.

eral falconry license. Even then, possession of eagle feathers is illegal, although the U.S. Fish and Wildlife Service collects eagle and other raptor feathers from permit holders for distribution to Native Americans for use in religious ceremonies.

Plucking Perches and Feather Rings

Raptors that feed on birds are fastidious about plucking their quarry before eating it, and during the nesting season, one or more sites may be used to prepare the food for their young. Such plucking perches or "butcher blocks" are rarely found in the case of the large falcons because of the general inaccessibility of their nest sites and environs, but they are a common feature near accipiter nests and are in fact another clue that a nest is nearby.

A plucking perch typically is uphill of the nest. The distance to the nest of a Sharp-shinned Hawk is usually very small—no more than about 17 m (50 ft) and usually much less. By contrast, a goshawk's plucking site may be 37 m (120 ft) or more away, although it is commonly closer than that. Stumps, horizontal branches not far off the ground, and the ground itself, most often under some overhanging branches, are all potential butcher blocks. Sometimes, an old nest is used, although it more commonly serves for temporary food storage.

A ring of feathers on the ground, .5 m (1 to 2 ft) in diameter, indicates a hawk has eaten here, and if out in the open, most likely a falcon. Additional proof is offered by the skeletal remains, if present: an intact pectoral girdle with the unplucked wings still attached and the sternum's keel notched by bites shows that your find was likely a falcon's kill. Sometimes, you may find two feather rings; after an initial going over, the falcon has dragged its prey a meter (1 yd) or so to keep plucked feathers from sticking to the exposed meat.

Where a Golden Eagle has made a meal of a large bird such as a heron, Mallard *(Anas platyrhynchos),* or Red-tailed Hawk, the scene is different: the feathers (including the large wing feathers) have been broken off and pulled out in bunches and scattered helter-skelter, sometimes over several meters, and leftovers, such as legs, may dangle from tree limbs, having been carried there in the final stages of feeding. It is a scene of great violence.

Behavior of Potential Prey

Birds and mammals that are the prey of raptors often indicate the presence of their predator by their behavior. Beechey Ground

Squirrels *(Spermophilus beecheyi)* whistle in alarm at the approach of a large raptor; if not in immediate danger, they stand erect and, their heads all pointed in the same direction, watch the passing of their nemesis. Such group staring can also be directed at a raptor already feeding on one of their comrades—or at a Coyote *(Canis latrans)* or Bobcat *(Felis rufus)*. If a group of ducks at the city park stares fixedly into the sky, following their gaze with binoculars may reveal a Peregrine Falcon overhead. A Black-tailed Deer *(Odocoileus hemionus)* was seen walking with oddly measured steps about a grassy field, her neck parallel to the ground, her head cocked skyward; not far above, a female Golden Eagle quartered back and forth, beak pointing down, searching for the doe's fawn concealed in the grass below (apparently, she had seen the fawn from afar before it hid; she did not find it but was herself driven away by an irate pair of territorial Red-tailed Hawks).

Providers of bird feeders are well acquainted with the mad dash for the bushes executed by their guests at the approach of a hawk (and with the many false alarms). European Starlings *(Sturnus vulgaris)*, blackbirds, and small shorebirds ball up in flocks, and then engage in aerial "dances" that are apparently intended to confuse the hawk. These swirling concentrated clouds of birds are an enthralling sight as they rise and drop in the sky, changing shape like windblown puffs of smoke. Scanning with binoculars usually reveals a nearby falcon or a high-flying accipiter: the Sharp-shinned Hawk, in particular, sometimes flies at great heights and pursues quarry there with falconlike stoops. The raucous alarm calls of a chorus of Steller's Jays *(Cyanocitta stelleri)* in the woods invite investigation; often, the object of their loathing is an owl that they have discovered among the leaves, but on occasion it turns out to be some diurnal raptor such as a Cooper's Hawk, glaring at its tormentors, perhaps with a defunct jay in its talons.

European Starlings are nearly ubiquitous these days and are only useful because of their warning calls. They have at least two kinds of alarm notes. One, a rasping squawk, indicates the presence of a terrestrial predator. It is given when a snake or squirrel climbs toward the starling's nest hole, for example. The other, a rapid-fire series of "tik-tik-tik" calls, alerts its neighbors to the proximity of an aerial predator, a hawk or a falcon. Interestingly, a starling flying over a Peregrine perched on a hill gave this call even though the predator was earthbound at the time. Once you have learned to recognize this call, it always pays to look up when you hear it. (A third call, a cluck, seems reserved for humans.)

When surprised out in the open, many birds freeze at the appearance of a hawk. An American Robin *(Turdus migratorius)* remains motionless for many minutes, and a Killdeer *(Charadrius vociferus)* leans forward and pushes its brown wing butts down and forward to conceal its brightly marked breast. A Willet *(Catoptrophorus semipalmatus)* at the sight of a falcon overhead raises its wings vertically and flashes the black and white wing pattern while giving alarm calls. Ducks often slip into the safety of the water from their shore resting places and usually do not fly unless severely pressed; a raptor, and especially a falcon, very rarely attacks ducks in the water. Herons and egrets in flight, worried by a raptor coming near, stretch out their previously folded necks to full length, perhaps to make themselves look larger, and utter croaking sounds. A White-faced Ibis *(Plegadis chihi)* plunged out of the air and landed in a rainwater pool at the approach of a low-flying Peregrine, which seemed to have no immediate interest in the wader. A nuthatch suddenly halts its mechanical scuttling on a tree limb or trunk and pretends it is just an old branch stub: a sure sign there is a Sharp-shinned Hawk nearby.

All of these prey responses are apparently predicated upon one or more subtle clues that indicate the raptor in sight is in fact hunting. Potential prey birds have an uncanny ability to "read" a hawk's intentions and react accordingly. Shorebirds, which rightfully regard a Peregrine as their ultimate nemesis, act fairly unconcerned if the passing falcon is not hunting or, more specifically, does not intend to attack *them*. The clue is not an obvious one, such as a visibly full crop, indicating that the bird has recently eaten; nor does it seem to be the speed of the falcon's wingbeat. Indeed, potential prey birds take wing under a falcon that specializes on some other species, an action that would ordinarily be suicidal.

How to Watch Hawks

Reading a Raptor's Mind

Hawks are as variable as people, some tolerating closer approach than others. The ability to determine a raptor's state of mind, and therefore predict its next move, can lead to better viewing. A hawk's beetling brows may give it a glowering appearance when in fact it is quite happy; more telling is its body language.

A hawk that sits upright, with one foot raised and partly hidden by the belly feathers and with its throat feathers puffed out, is a hawk at ease. Importantly, the hawk does not watch the observer, an indication that the latter has not approached too close for the raptor's liking. In fact, it may partly or completely close an eye, though usually keeping the other open. Eventually, the hawk may lower its foot and rouse, or it may stretch its wings, singly or together. Should the hawk be feeding or preening, it leisurely continues to do so. Observers should not approach so closely that a raptor feels pressed to leave its hard-won meal; in some species, such deprivation can be literally a matter of life or death.

By contrast, a nervous hawk stands on both feet; its appearance is sleek regardless of the weather, and its body position diagonal or horizontal. It may stretch its neck or bob its head up and down, peering at the observer. If it is feeding, it will stop doing so. These are all signals of the bird's imminent departure; as a final sign, the hawk raises its tail and defecates.

Should a wild hawk suddenly mantle while feeding, it has spotted another approaching raptor that might take its prey, and it may pay to ready your camera for a spectacular photo. If it suddenly bolts, leaving its quarry, it is fleeing a much larger, dangerous raptor.

Binoculars and Spotting Scopes

No matter how tame the hawk, binoculars are the sine qua non of raptor watching. Even the cheapest and most humble pair of opera glasses is better than none at all. However, expensive binoculars are much better than cheap ones when it comes to sharpness of image across the entire field of vision, sturdiness, and especially the ability to gather light under low-light conditions, making it possible to still see much detail.

There are two basic kinds of binoculars, the porro prism type and the roof prism variety. Porro prisms have thick barrels, with the much smaller eyepieces offset from the center of the barrels; the best are very highly rated in terms of image sharpness and light gathering, but all suffer from a certain delicacy that causes them to lose their alignment after seemingly minor impacts. Roof prisms have straight barrels, in line with the eyepieces; their compact shape, smaller size, and sturdiness more than compensate for their supposedly slightly inferior performance in poor light, and they have become the favorite of most advanced birders.

When choosing binoculars, weight is a major consideration. High-powered binoculars, such as 12 × 50, can be very heavy and difficult to carry for long periods or hold steady while viewing a bird; 8 × 40 or 10 × 40 binoculars are generally more practical, although their magnification is a bit less.

Spotting scopes are designed for serious birders who do not mind carrying with them a tripod, a virtual must for this instrument. Spotting scopes are wonderful for really visiting with a wild hawk; because of their great magnifying power, they can be used from several hundred meters away, and they can reveal extraordinary detail. Magnification ranges from 10× to 250×, but the higher the magnification, the dimmer the image and the narrower the depth of field and field of view. Spotting scopes are ideal for observing the bands on the legs of some raptors (and in some cases even reading the numbers on these bands) and watching the young in a distant eyrie; mounted on a car window, they are also useful for raptor watching along the road. Practice is required for watching hawks in flight. Most modern spotting scopes can be fitted to cameras and used as telephoto lenses as well, although the resulting pictures are not always high quality.

Buying a good scope or pair of binoculars can be a major investment, and it pays to obtain the opinions of regular users of a desired model; Audubon Society members and other outdoor people are usually happy to share such information.

Cameras and Photography

Taking pictures of wildlife has become a minor industry, and among the enthusiasts are a substantial number that specialize in raptors. Modern cameras typically have not only several kinds of built-in light meters but also such features as automatic selection of the fastest shutter speed under available light (which makes it easier to get good pictures of fast-moving hawks) or of the aperture for certain artistic effects; more important, perhaps, are the automatic film advance, automatic focus, and rapid-fire exposure options. Automatic focusing works well for raptors perched in the open or flying close by. Unfortunately, autofocus also has an irritating tendency to lock onto twiggery near or in front of a hawk, forcing the photographer to resort to manual focusing. Predictive autofocus keeps a bird in focus as it flies in the general direction of the photographer.

Single lens reflex (SLR) cameras with these features and fitted with various lenses have been the standard for wildlife photography. Recently, high-end digital SLR cameras that accept the long lenses of film cameras have become available. Magazines published by and for bird-watchers occasionally feature articles covering the latest advances in equipment.

Photographing Captive Raptors

Even very tame raptors often react nervously or aggressively to a very close camera lens; perhaps they perceive it as a predator's eye. A medium telephoto lens such as a 135- or 200-mm lens, which allows the photographer to stand back at least several meters and still take a close-up shot, produces the best photos.

Taking pictures of captive hawks can yield very satisfactory results provided you do not try to claim that they are pictures of wild raptors. Birds of prey kept in zoos or rehab centers very often have grossly overgrown beaks and talons, the result of an unnatural diet and environment. If a hawk portrait shows an indentation of the feathers where head and neck meet, or if a tuft of feathers sticks out from the nape, the bird has just had its falconer's hood removed. Raw areas at the gape-cere junction are evidence that the hawk often wears an ill-fitting hood.

Commonly, hoping to pass off a photo as that of a wild hawk, a photographer makes a clumsy effort to hide the hawk's legs behind a branch or grass clump to hide the telltale cuffs and jesses (the leather bracelets and straps that captive hawks wear as restraints). Also, wild hawks rarely have a broken feather, and their plumage is in disarray for only very brief periods. A hawk fiercely mantling over prey, photographed with its beak wide open in midscream is almost certainly a captive, imprinted as a chick on a human. Odd highlights in the eye are also a giveaway, as are the hunched "shoulders" and half-open beaks of frightened raptors (usually wild ones having just been caught) held by an unseen hand. A new picture book of raptor photos or bird art presents the challenge and fun of figuring out which portraits of supposedly wild birds are actually of captives and which paintings by wildlife artists were copied from photographs of such birds.

Fig. 48. A Merlin wearing a falconry hood and jesses.

Fig. 49. A Peregrine Falcon with hood just removed. The indentation left from the hood (at arrow, often more conspicuous) breaks the malar stripe. Note the somewhat overgrown beak.

Photographing Wild Raptors

Photographing wild raptors ordinarily excludes most shenanigans; the only one that is at all common is the placement of a tethered prey animal within sight of the raptor so that predation can be photographed at a predetermined time and place. Alas, gerbils, which are Old World desert rodents commonly sold as pets, are sometimes used as bait and are easily recognizable, particularly when placed on snow.

The easiest way to get within camera range of a wild raptor is to drive up to it in a car, which becomes a de facto movable blind. If you must go on foot, it helps to approach the bird on a tangent, rather than in a direct line, as if intending to walk past it while giving it some berth.

No matter how close you can get to a wild hawk, photographing it nearly always requires a sizeable telephoto lens. Zoom lenses are generally less desirable because they require more light than is often available, and they frequently are not powerful enough. A 400-mm, f5.6 lens, which with practice and the use of fast film can be hand held, is adequate though not ideal, especially in low light.

A 500-mm or especially a 1000-mm lens is far superior but necessitates the use of a sturdy and hence heavy tripod that is not likely to rock in the wind. Lower aperture numbers in long lenses, such as f2.8, although permitting faster shutter speeds and photography under low-light conditions, unfortunately add thousands of dollars to the cost of a good lens.

The splendid images of raptors in magazines are the very best selected by the photographer from a great number of exposed rolls. Outstanding raptor pictures are the result of countless hours in the field and the willingness to greatly enrich the makers of film.

Photos taken at a hawk's nest can be spectacular, but unless done from afar, authorities generally frown on such photography, especially in the case of species enjoying special protection. Only urban raptor nests are likely sites, provided the adult hawks are habituated to the presence of humans; even here, possessive humans living nearby often do not like to see "their" hawks disturbed, and rise in protest. Construction of blinds in which the photographer can hide for many cramped hours facilitates taking pictures at nests.

Hawk-watching Etiquette

The increasing popularity of watching hawks has its down side. The presence of people can disrupt the daily activities of hawks by interrupting a chase, driving a hawk off its kill or making part of a hawk's territory temporarily unusable. Nesting raptors may abandon a nest if people visit the nest area too often or for too long a time, especially during the critical period of early incubation. An area holding an accidentally discovered nest should be left expeditiously.

Some raptors, such as Cooper's and Sharp-shinned Hawks, can be attracted by "squeaking" (imitating an injured prey animal). Mimicking the call of a Great Horned Owl *(Bubo virginianus)* sometimes works also. Unfortunately, it is easy to overdo both squeaking and hooting so that the hawks become habituated to the sounds and ignore them, thereby losing legitimate hunting opportunities; some individuals may leave the area altogether.

Photographers frequently make several attempts to get close to a hawk, especially one that is slowed by a large prey item, only to find that with each try, the bird's flight distance gets larger, thereby defeating the photographer's purpose. The hawk may even ultimately abandon its meal. One photographer staked out for a day a solitary oak known as an often-used perch of Golden Eagles; he was unaware that no more than 30 yd behind his parked car, in a grove of live oaks, a female Golden Eagle was incubating two newly laid eggs. She abandoned the nest.

A comprehensive list of sensible rules for watching all birds can be found on the Web at americanbirding.org/abaethics.htm.

Attitudes Past and Present

I often feed small birds, but when they come together,
hawks spot them and catch them—a very bad thing. So
in order to protect these small birds, I keep the air rifle.

—THE DALAI LAMA
(FROM AN INTERVIEW IN THE *SAN FRANCISCO CHRONICLE,*
DECEMBER 5, 1993) ©1993, *THE NEW YORK TIMES MAGAZINE*

In the year 2000 in the Asian Republic of Georgia, you could dine on boiled and roasted hawks that had been intercepted on their southbound migration, and no doubt you can do so today; in a number of countries, thousands of raptors are shot annually for amusement and discarded or stuffed. In Pennsylvania, where there are famous hawk migration bottlenecks, "bad" raptors were slaughtered legally until the 1930s. Californians have never en-gaged in such activities on a grand scale, perhaps in part because hawk funnels such as the Marin Headlands are relatively recent discoveries. But birds of prey have not always enjoyed unalloyed popularity in our state. Until the late 1960s, even many bird-watchers (including some famous ornithologists who should have known better) disliked hawks for snatching songbirds off their backyard feeders, and one prominent scientist decided that some of California's last surviving Peregrines *(Falco peregrinus),* then critically endangered, should become museum specimens. Hunters and ranchers routinely shot the hawks they encoun-tered—the latter hanging them in rows from fences. Today in California, however, shooting hawks is illegal and rare, and most enlightened people, with some astonishing exceptions (see the quote above), are aware of the beneficial interconnections be-tween predators and prey, and many of us are thrilled to watch a hawk go about its natural business.

The first California hawk-watchers—the indigenous Califor-nians—ascribed sacred and magical properties to raptors, which, however, also had utilitarian merits. Raptors are featured in the songs and stories of many tribes, and their feathers were com-monly used for arrow fletching. Condor bones were fashioned into flutes.

California Indians were as spiritually connected to eagles, as were the Indians of the Great Plains and Southwest; eagles

Fig. 50. Tip of a Maidu wooden ceremonial hairpin, adorned with a male American Kestrel's tail feathers, topknot feathers of a California Quail *(Callipepla californica),* and the scalp of a woodpecker.

figured prominently in their stories and creation mythology, and clothing and headgear were decorated with eagle feathers as well as with those of Cooper's Hawks *(Accipiter cooperii),* Red-tailed Hawks *(Buteo jamaicensis),* American Kestrels *(Falco sparverius),* Peregrine Falcons, and occasionally other species. California Condor *(Gymnogyps californianus)* and Golden Eagle *(Aquila chrysaetos)* feathers (especially the striking bicolored tail feathers of first-year eagles) were valued for making ceremonial cloaks. The pouches of medicine men often contained a hawk's foot or an eagle skull, and sometimes even a whole stuffed raptor. Birds of prey in general were considered power symbols. In southern California, they were deities (Heizer 1978). The Chumash of Santa Barbara County attributed eclipses to the passing wings of a mythological Golden Eagle, and Condors were believed to help people locate lost objects (Timbrook and Johnson 1999). "Condor Cave," a winter solstice observatory of the Chumash, contains a colored pictograph that is unmistakably a Condor.

The California Condor held special significance. Archeological digs along San Francisco Bay unearthed Condors buried, apparently with great respect, in shell mounds. Many tribes had traditions that required Condors to be sacrificed, often to honor deceased leaders or transfer the bird's power or other attributes to the participants. The Luiseños, among other tribes, held annual celebrations during which a Condor was killed without drawing blood. In Condor impersonation ceremonies, dancers wore Condor skins and mimicked the bird's hiss and movements. Miwok shamans, for example, obtained supernatural curative powers in this way (Snyder and Snyder 2000).

For some ceremonies, the more numerous Golden Eagle or a buteo was used when a Condor could not be acquired or when celestial events demanded a different species. The birds were either trapped as adults or taken from the nest and raised in the villages. Known Condor nests were owned by the tribe. Birds not killed in a ceremony were sometimes sold to another tribe or village unable to obtain one otherwise for its own rituals. Eagles were also trapped for food. (Kroeber 1953).

Most early anthropologists were not biologists, and even prominent students of California Indians in the early twentieth century were poorly trained in zoology and botany. When it came to wildlife, they tended to write about the most conspicuous species, and they often got their information wrong. Mentions of the less spectacular raptors are uncommon, either because the scientists had less interest in them or perhaps because they had less significance to the people they were studying.

Memoirs of early European settlers and explorers contain tantalizing but scant accounts of their encounters with "very tame" hawks, some of which were eaten by the desperately hungry travelers. The gold miners who followed them used Condor quills as gold-dust containers that sold for a dollar apiece.

Conservation Laws and History

Because birds of prey were believed to compete with hunters and to rob poultry farmers of their profits, all raptors were once widely regarded as vermin. Attitudes started to slowly change, however, after the 1893 publication by the U.S. Department of Agriculture of a study that recorded the stomach contents of birds of prey. The report showed that most hawks, being rodent eaters, were in fact economically beneficial to agriculture, and many states started to pass laws protecting those species.

The bird-eating hawks, however, were labeled as definitely harmful, and for half a century all states with the exception of Hawaii sanctioned the killing of at least the accipiters (with some states even offering a bounty) along with other raptors and birds deemed harmful, such as the White Pelican *(Pelecanus erythrorhynchos)* and Black-billed Magpie *(Pica pica)*. California enacted a law to protect most raptors in 1901 (Mallette and Gould 1976) but allowed the shooting of Cooper's Hawks, Sharp-shinned Hawks *(Accipiter striatus),* and Peregrine Falcons (Sprunt 1955). Participants of a jay shoot organized by the Associated Sportsmen of Calaveras County in 1936 were awarded one point for each jay shot and three points for any hawk (Erickson 1937). The California Condor has had special protection since before 1890, but that state law was ineffective in part because it was so little known.

Federal protection of birds of prey came more slowly. Raptors were in fact omitted from the first major federal action to protect wild birds, the Migratory Bird Treaty between the United States and Britain (for Canada), which was implemented, in part to end the feather trade for women's hats, in 1918. However, because the treaty outlawed not only the killing or taking of most other birds but also their eggs, it effectively ended the popularity of egg collecting, a hobby that eventually could have had an impact on raptor species in particular because of their often attractive eggs and low population numbers and reproductive rates. Subsequent amendments, together with additional international treaties, extended protection to all birds of prey (eagles, hawks, and owls were added in 1972). The federal laws took precedence over state regulations and resulted in better public awareness of the law and stiffer law enforcement.

Other important federal legislation that benefits California's raptors includes the banning of the pesticide DDT in 1972 and the federal Endangered Species Act of 1973 that expanded a 1966 law. California passed its own Endangered Species Act in 1970; unlike the federal act, however, it provides protection for species on state-owned land only. Funding problems and politics often limit the goals of both federal and state laws.

Although the public perception of endangered species acts is one of powerful regulations that often disallow the development of private lands, in reality this rarely happens. A developer may mitigate the "taking" (i.e., disturbance or destruction) of listed species or their habitat by preserving part of the parcel for the species, donating suitable habitat for the species in another area, restoring suitable damaged habitat elsewhere, relocating the animals, or funding efforts to preserve other listed species.

The federal Bald Eagle Protection Act of 1940, which was written specifically to protect Bald Eagles *(Haliaeetus leucocephalus)* and in 1962 was amended to include Golden Eagles, is stronger than the Migratory Bird Treaty Act because it includes prohibitions even against disturbing the birds and prevents the taking of eagle nests (and, therefore, nest trees in the path of development). However, in a nod to mining interests, the act was amended to allow the "taking" of Golden Eagle nests by permit providing that the nest is "inactive," (meaning any time other than when an eagle, its eggs, or young are on or in the nest); real estate developers, of course, also have utilized this clause. That the same nest may have been used for decades is of no concern. Mitigation is merely "encouraged." Theoretically, issuing a "take" permit is predicated upon the health of the local eagle population.

California's Threatened and Endangered Species

California Condor

The California Condor was listed as federally endangered in 1967; in California, it was listed as endangered in 1971.

Fossil records indicate that the California Condor was once widely distributed over much of what is now the United States. The disappearance of mammoths and other large mammals after the arrival of humans from Asia deprived them of their principal food source and reduced their range. Early accounts of Condors by Europeans indicate that the species could be found up and down the Pacific coast from British Columbia into Baja California, where washed-up whales, sea lions, and the Northern Elephant Seal *(Mirounga angustirostris)* remained abundant. Recent isotopic analyses of parts of old museum specimens have confirmed that marine organisms, most surely large marine mammals, were an important part of their diet at that time. The decimation of the large congregations of sea lions and elephant seals to supply oil for heating and lighting to the expanding human populations and the near extinction of the large whales for the same purpose significantly reduced the quantities of this food supply. With the arrival of the gold-seeking Forty-niners, available food in the interior also decreased as elk (*Cervus elaphus* ssp.), Pronghorn *(Antilocapra americana),* and Bighorn Sheep *(Ovis canadensis)* were shot for food and sport. The development of large cattle ranches and the year-round availability of carcasses of horses and other domestic livestock permitted the survival of condors throughout most of the twentieth century in southern California, although in ever decreasing numbers.

Condors made attractive targets; one ornithologist reported, "in the late 1800s it was the ambition of every boy to shoot a condor" (Koford 1953). But Condors were facing other more serious threats; their numbers were declining dramatically, probably in part from feeding on poisoned carcasses put out by cattle ranchers to kill predators (including Grizzly Bears *[Ursus arctos]* and Gray Wolves *[Canis lupus]*). The Condor population decline was so noticeable that in 1937 Congress established a refuge for the birds, the Sisquoc Condor Sanctuary in Santa Barbara County, the first of many actions taken over the following decades aimed at protecting Condors and their habitat. A second sanctuary, the Sespe in Ventura County, was created in 1947. Condors, however, range outside sanctuary boundaries; their numbers continued to fall.

The Condor conservation story is a fascinating study of controversy, political and legal maneuvering, infighting, and self-interest; a battle over captive breeding raged until only eight birds (including just one breeding pair) remained, and the decision was made to remove them from the wild. The loss of wild Condors, some feared, might jeopardize funding for Condor programs and limit the ability to acquire Condor habitat. The Chumash argued that the plan would interfere with their religion (Snyder and Snyder 2000). Nevertheless, in 1987, after a final court fight, the last wild California Condor was captured.

All extant California Condors are descended from just 14 individuals. Captive breeding supplied Condors that were released to the wild beginning in 1992. Three eggs laid in the wild in California in 2002 resulted in the first wild-produced young since 1984. All three chicks, however, as well as the single California chick hatched in 2003, died near fledging, some of them, perhaps, from the numerous bottle caps and other artifacts found in their crops and stomachs. The first wild chick to survive past fledging was hatched in Arizona in 2003 and is doing well (as of 2004). The Los Angeles Zoo, the San Diego Wild Animal Park, and facilities in Idaho and Oregon produce Condor chicks for release by the Ventana Wilderness Society, the United States Park Service at Pinnacles National Monument, the United States Fish and Wildlife Service Hopper Mountain National Wildlife Refuge, The Peregrine Fund (Arizona releases), and the Zoological Society of San Diego (Baja California releases).

Environmental hazards threaten the long-term success of the Condor as a free-living, independent species. Condors die in collisions with overhead wires. Some have died from drowning, and others have been electrocuted on high-tension towers. One bird died from ingesting antifreeze.

Sick Condors have been recaptured, treated for lead poisoning, and rereleased; lead is ingested when the Condors feed on the carcasses or discarded parts of animals shot by hunters. A single shotgun pellet can kill a Condor. Although not documented as a major cause of death of wild Condors until the 1980s, lead may have played a role in the species' decline for a century or two (Snyder and Snyder 2000) and in 2003 was tied with death from being

Fig. 51. The Sespe Condor Sanctuary (Ventura County) is located in traditional Condor nesting habitat.

attacked by Golden Eagles (at five birds each) as the leading cause of mortality of released Condors (L. Kiff, pers. comm. 2003). Wild-raised young may also be at risk as lead fragments are passed to them by their parents during feeding.

A partial solution to the lead problem may be to require that only nonlead bullets be used in areas where Condors live, just as duck hunters are required to use nonlead shot over water; before this restriction was enacted, lead poisoning was a leading cause of Bald Eagle mortality. Laws and regulations enacted to protect the Condor include restrictions of people, firearms, poisons, and air traffic in sensitive areas, yet a roadless area in Los Padres National Forest where two of the Condor eggs were laid in 2002 has been considered for oil drilling.

Some captive-bred Condors, once released, are particularly casual about the presence of humans, and actually show an attraction to human-built structures, conduct that has led to problems including vandalism of garden hoses and other human artifacts by the young Condors. Birds that approach humans or otherwise misbehave are returned to their release pen before others begin to copy their bad behavior; a year in Condor jail usually reforms them. Young birds tend to fare better when raised by adult Condors (and not fed by humans) and released with formerly free-living birds (those captured in the 1980s) that can demonstrate proper Condor behavior and keep the youngsters out of trouble (Snyder and Snyder 2000). In 2003, this small pool of mentor birds was reduced by one, which was shot and killed by a hog hunter not quite three years after the bird's return to the wild.

Bald Eagle

The Bald Eagle was listed as federally endangered in 1967, but the status of this species was reduced to threatened in 1995, and it was proposed for delisting in 1999. In California, it has been listed as endangered since 1971.

Bald Eagles were widespread in California before the Gold Rush, when their numbers began to decline for some of the same reasons (shooting and poisoning) that those of the Condor declined. Early settlers throughout North America mistakenly believed that Bald Eagles killed young livestock and carried off human babies. Many were shot during their migrations southward from Canada and Alaska, and a reported 128,000 were killed for a bounty offered by Alaska as late as 1953. The Bald Eagle Protection Act of 1940 made it illegal to kill the birds elsewhere in U.S. territory, but shooting and electrocution are responsible for the majority of Bald Eagle deaths in California today. Some ranchers still suggest with a smile that all eagles can die from "lead poisoning"—in one of two ways.

Bald Eagles feed mostly on fish and waterfowl but also eat carrion, and poisoned carcasses put out for predator control have killed many. However, the most significant reason for their decline starting in the 1940s was the effect of a breakdown product (DDE) of the pesticide DDT (Garcelon et al. 1989), which causes birds to produce eggs with shells too porous to prevent water loss. DDE accumulates in the fat deposits of animals; when an animal is eaten, its DDE is transferred to the consumer. A fish that feeds on smaller fish that have, in turn, fed on predatory invertebrates, will have accumulated the toxins stored in all the animals along the food chain by the time it is eaten by a Bald Eagle. Raptors that prey primarily on plant eaters, such as rodents, are less affected by DDE, because they take in lower levels of the toxin than those at the end of longer food chains.

Bald Eagle numbers were at their lowest levels in 1972, when

DDT was banned for most uses in the United States. At that time, the species had stopped breeding in the southern part of California. Now, Bald Eagle numbers are climbing, but DDT is persistent in the environment and continues to affect some eagles' ability to reproduce. From the late 1940s through 1970, a DDT factory in southern Los Angeles County dumped or discharged a minimal estimate of 2,000 metric tons of DDT compounds into the ocean off the southern Los Angeles County coast (R. Risebrough, pers. comm. 2003). Bald Eagles reintroduced to Santa Catalina Island (the nearest to the outfall from the factory of all the Channel Islands, where the entire natural population died out) as of 2003 continue to produce defective eggs. Therefore, all eggs are removed from the nests and artificially incubated at the San Francisco Zoo. A few eagle pairs, however, appear to be producing some eggs that require less care and may eventually raise young without human manipulation (D. Garcelon, pers. comm. 2003). The Institute for Wildlife Studies is repopulating Santa Cruz Island with eaglets from Alaska in a study to determine whether the northern Channel Islands are as polluted.

The recovery of the Bald Eagle has been aided by governmental agencies monitoring populations, saving foraging habitat, and writing timber guidelines to protect nesting eagles and even individual trees used by eagles for nesting or roosting. The Nature Conservancy and the National Wildlife Federation have purchased historic roosting sites. Habitat protection, however, remains a critical concern. The Ventana Wilderness Society at Big Sur also maintains a successful captive-breeding program, and released birds are reproducing in the wild.

Swainson's Hawk

The Swainson's Hawk *(Buteo swainsoni)* is not listed federally, but in California, it was listed as threatened in 1983.

California provides breeding habitat for Swainson's Hawks, which used to be so common throughout the grassy lowlands of the state that early naturalists thought them hardly worth mentioning (Bloom 1980). In some areas the species was the most abundant breeding hawk (Sharp 1902), although the same author reported it scarce prior to 1897. Between the early 1900s and 1979, however, the breeding population statewide dropped to less than 10 percent of its historic size (Bloom 1980). Of California's raptors, only the Condor has experienced a greater decline. Although several factors may have contributed, the principal cause simply is not known, but there are several possibilities.

Habitat destruction—the removal or fragmentation of the woodlands and riparian habitats where the species nests and loss of foraging habitat both in nesting territories and along the migration route in Mexico—may have played a role. One study estimated a 98 percent reduction of riparian habitat in California's Central Valley (a prime nesting area for the hawks) since presettlement times (Schlorff and Bloom 1983). Flood control and riverbank protection projects destroy habitat as does the unabated conversion of the hawks' hunting fields (chiefly grasslands, pastures, and low-density vegetation croplands where prey is plentiful and accessible) to vineyards, orchards, housing, and shopping centers. However, local populations also have declined in areas of California where sufficient habitat still exists (such as in the central Coast Ranges, various southern California regions, and the Owens Valley), indicating some other cause (Risebrough et al. 1989).

Swainson's Hawks make easy targets because of their tameness, their often conspicuous nests, and their habit of roosting together in very large numbers (sometimes even on the ground) and failing to fly when others around them are shot. Reportedly, groups of many hundreds are sometimes shot at South American winter roosts (Bloom 1980).

A report of 700 wintering Swainson's Hawks found dead after the single application of a pesticide to one field in Argentina raised suspicions that poorly regulated pesticide use in other countries may be a factor in the population decline in California; one researcher discovered a bird banded at a nest in his own northern California study area amid a pile of moldering Swainson's Hawk corpses under an Argentine roost site (Woodbridge, Finley, and Seager 1995). However, most North American breeding populations of the Swainson's Hawk outside California (which also winter in South America) have not experienced marked declines and are not considered threatened.

Peregrine Falcon

The Peregrine Falcon was listed as federally endangered in 1970 and delisted in 1999. In California, it has been listed as endangered since 1971.

Like the Bald Eagle, the Peregrine Falcon was severely affected by DDT/DDE throughout North America and much of Europe as well. By 1970, the species had stopped breeding completely in the eastern United States and had disappeared; California had only a few known breeding pairs, down from an estimated 150 nesting pairs pre-DDT (B. Walton, pers. comm. 2002). In this species, the pesticide caused the birds to lay thin-shelled eggs that broke during incubation or were otherwise defective. Shooting was also a major threat, and continued to be so after DDT was banned.

In 1977, the Predatory Bird Research Group at the University of California at Santa Cruz set out to restore the Peregrine population in the west through captive breeding and release to the wild. Based on the successful Peregrine reintroduction program begun a few years earlier on the east coast by The Peregrine Fund, the group's efforts greatly augmented the recovering population, although it is likely that the population would have recovered on its own, albeit much more slowly, from remaining populations of Peregrines in Arizona and Baja California.

Although the Peregrine Falcon (Anatum race) is still listed as endangered in the state, it is a remarkable example of population collapse and recovery. There are almost certainly more Peregrines today in California than there were in the 1940s (before DDT); the number of estimated nesting pairs at present is 250 and growing (B. Walton, pers. comm. 2002). The fear that habitat destruction (especially development in coastal southern California near historic cliff-nesting sites) would threaten their success has proved unfounded, for Peregrines are now making use of human-made structures (such as bridges and high-rises), nest sites that were previously not recorded in California.

California's Species of Special Concern

COOPER'S HAWK *(Accipiter cooperii)* AND SHARP-SHINNED HAWK *(A. striatus)*: Migration counts show substantial number fluctuations in these bird-eating species. Breeding populations declined in the 1950s and 1960s, but lack of up-to-date figures leaves their current status in doubt; they are probably more numerous than officially believed.

NORTHERN GOSHAWK *(Accipiter gentilis)*: California's small breeding population of 600 to 800 pairs (J. Keane, pers. comm. 2002) has suffered from habitat destruction by the timber industry.

GOLDEN EAGLE *(Aquila chrysaetos):* In contrast to Golden Eagle numbers in southern California, those in the center and north of the state appear to be mostly holding their own despite accelerating habitat loss and an annual kill of between 30 and 70 eagles by wind turbines in the Altamont Pass, Alameda County. The eagles destroyed here are chiefly subadults and floaters, that is, adults that have no nesting territories, indicating a saturation of available suitable territories.

FERRUGINOUS HAWK *(Buteo regalis):* Wintering populations are decreasing throughout its range in California mostly because of agricultural development, rodent control, and perhaps shooting.

NORTHERN HARRIER *(Circus cyaneus):* The preference of harriers for grasslands and marshes is also their downfall because of the conversion of such areas into housing tracts; not only is there an accelerated loss of habitat but also the introduction of cats and dogs has led to the destruction of the prey and the nests of these hawks.

OSPREY *(Pandion haliaeetus):* This species presumably suffered from DDT but has recovered nicely, especially in northern California. Populations in southern California declined from unknown causes before pesticides became a problem, although in some areas such as the Channel Islands, indiscriminate shooting is suspected (Kiff 1980).

HARRIS'S HAWK *(Parabuteo unicinctus):* Fairly recently extirpated as a breeding species in California, this desert hawk is being reintroduced. However, it has been suggested that Harris's Hawks may not actually be a normal part of California's breeding raptor fauna.

MERLIN *(Falco columbarius):* This species is a winter visitor/transient only; reproductive failure from DDT may have caused a population decline. Today, wintering Merlins are more common than generally realized through much of the state.

PRAIRIE FALCON *(Falco mexicanus):* Breeding numbers seem in decline for unknown reasons, though wintering birds are widespread.

Conservation Methods

In the past, conservation has been focused on saving individual species, particularly charismatic ones with which the public can identify. Finding funding, both public and private, to protect the Bald Eagle is not as difficult as generating interest in saving a less well-known raptor or a plant or insect. The single-species approach, however, is increasingly seen to have too small a focus. To better preserve all species (not just those that are already in crisis), the emphasis is shifting to the protection of whole ecologic communities (such as an entire watershed), as well as preserving habitat along migration routes within the United States and encouraging the same where those routes stretch beyond our borders. Migration routes, along with nesting and wintering grounds, must offer adequate food, and all areas need to be free of hazards.

Habitat protection, which can involve the purchase, restoration, or both of sensitive areas, is often an important part of recovery plans. Restoration may require the elimination of alien species and the reintroduction of native species (whether grasses or eagles) that have disappeared from an area.

Sometimes conflicts arise. The burgeoning Peregrine population in California has run afoul of efforts to restore the endangered California Least Tern *(Sterna antillarum browni)*, resulting in a relocation of falcon chicks, and the same offense resulted in the shooting of Northern Harriers at Camp Pendleton in San Luis Obispo County. Golden Eagles, recently (and naturally) arrived on the Channel Islands in southern California, are being trapped and relocated (although the eagle is a California Species of Special Concern) because they prey on the more threatened Island Fox *(Urocyon littoralis)*, which lives nowhere else. Recently Bald Eagles, which were pre-DDT inhabitants of the islands and had kept out the Goldens but are no danger to the fox, have been reintroduced. Forest restoration plans (to aid the recovery of Northern Goshawks, for example) may include prescribed fire and forest thinning in an attempt to recreate a more natural (pre-fire suppression) habitat, but the idea remains controversial.

Captive-breeding and reintroduction programs have increased the populations of some species; success depends on an adequate food supply for released birds, the lack of environmental hazards in their habitats, adequate nesting areas, and, of course, the quality of the program and the skill of the people involved. Reports in the news of Bald Eagles coming down to "attack" little girls or of California Condors vandalizing buildings can be attributed to captive-bred birds that had become habituated to humans while in captivity. Some problem birds have been recaptured and given human-avoidance training, with limited success; ideally, birds raised for future release should have no visual contact with people.

Several clever techniques have been developed to increase production of captive-raised raptors. Double clutching produces double the normal number of offspring; a bird whose eggs are removed from the nest will lay another set, and the first set can be hatched in an incubator or placed in the nest of a nonthreatened compatible species (for example, young Peregrines have been raised in the wild by Prairie Falcons). Because the first eggs produced in the wild by released Condors in 2001 did not survive, scientists replaced them first with a ceramic fake and then, to give the parents experience for a following year, with a ready-to-hatch Condor egg produced in captivity.

A variety of raptors can be enticed to use human-made nest sites. American Kestrels readily accept nest boxes placed in appropriate habitat, Peregrine Falcons lay their eggs in large trays with gravel floors on building ledges, and Ospreys, Bald Eagles, and even occasionally Golden Eagles may build nests on artificial platforms.

The most common method of releasing captive-raised raptors to the wild is by hacking. Young birds are kept and fed in a "hack box" (a wooden crate, large enough to allow the youngsters to exercise their wings, with barred windows) at a release site; they familiarize themselves with the location as they mature. The box is then opened, and they are free to leave. Normally, the birds remain in the area to practice their flying skills and eat the food that is still left for them in the box before they learn to hunt sufficiently for themselves and disperse. As adults, the birds tend to return to nest in the areas where they were released.

Disturbance by Humans

Without question, human alteration of the land has had a significant effect on California's raptors. For some species, some forms of human disturbance can lead to increased populations. White-tailed Kites *(Elanus leucurus)* seem to benefit from moderate habitat disturbance by humans and from human-made habitats, as long as a prey base and nest sites remain, and so do

Fig. 52. A White-tailed Kite hovers over a remnant grassland fragment adjacent to a new subdivision.

Fig. 53. American Kestrel, juvenile male, struck and killed by a car.

Red-tailed Hawks. Land converted to pasture may be good hunting habitat for Swainson's Hawks, and in some areas Ferruginous Hawks prefer heavily grazed lands for foraging, because prey is more visible on bare ground (Wakeley 1978). When sagebrush takes over where native grasses are eaten out, however, prey be-

Fig. 54. A raptor fed on this ground squirrel, which was killed with grain bait laced with an anticoagulant. The turquoise tracing dye that coats the poisonous bait binds to the squirrel's subcutaneous fat (the skin was turned inside out by the raptor).

comes less vulnerable and less abundant. In general, development decreases diversity. As habitat is lost to urban sprawl and industrial parks, displaced individuals have no guarantee of finding open territories in which to take refuge. Increased traffic is a peril to those species (such as Red-tailed Hawks and American Kestrels) that use the narrow strips of vegetation next to roads and freeways for hunting.

Although clear-cutting of mountain timber receives much press, the steady loss of California's oak woodlands, which affects a greater number of wildlife species (including many of California's raptors), is scarcely noticed. The oaks, which provide nest sites, roosting perches, and food and shelter for prey species, are cut for firewood and to make room for orchards, vineyards, pasture, and housing. Many of our large oaks are reaching the end of their 200- to 300-year life spans. Most young (replacement) oaks are found growing outside pasture fences, where cattle cannot reach them, and ranchers and others are becoming aware that grazing reforms are needed.

Toxins are a continuing threat to raptors. Turkey Vultures

Fig. 55. Turkey Vultures feeding on poisoned squirrel (blue dye visible).

(Cathartes aura) are virtually immune to botulism; surprisingly, neither are they very susceptible to the anticoagulants in squirrel poisons, which kill Golden Eagles that feed on the dying squirrels (Peeters 1994). Each year an average of 43 tons of poisoned grain are distributed in Alameda County alone (J. Gouvaia, pers. comm. 2002), most of it to cattle ranchers to kill ground squirrels. Ironically, overgrazing by cattle encourages the squirrels to colonize the land, because squirrels dislike tall, dense grass. The introduction of an alien species (such as cattle) nearly always causes ecological disruptions.

Many agricultural pesticides besides DDT are harmful to hawks, and some can kill them outright. Even when certain chemicals are banned for use in the United States, migrating hawks may pick up toxins on their tropical wintering grounds, or migrating prey species may return to California with high levels. Some Latin American countries impose no limits on duck hunters, so fewer ducks return to California, and those that do may carry lethal lead shot. On the other hand, we often forget that our actions can affect other countries. When Bald Eagle populations plummeted in the United States, the numbers of breeding pairs in Mexico also declined. After the U.S. ban of DDT, Mexican populations, never large, began to recover (G. Ceballos, pers. comm. 2001) although DDT continued to be used in Mexico. DDT use is now greatly reduced in Mexico and banned in most Latin American countries (L. Kiff, pers. comm. 2003).

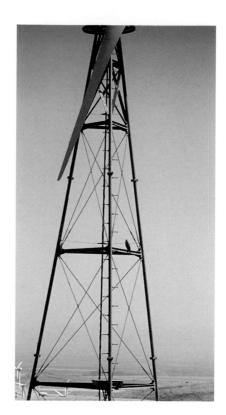

Fig. 56. Red-tailed Hawk perched on a wind turbine.

The Problems with Wind Power

Wind power has been highly touted in the past two decades as a clean energy source, and wind farms appear to have little impact on the fauna of an area. The one exception is the installation of thousands of turbines in the eastern part of Alameda County, where every year the blades kill hundreds of raptors. The hills in this area form a wind funnel that greatly facilitates soaring, and they are rich in ground squirrels, thus attracting large numbers of birds of prey. Chopped up Golden Eagles and Red-tailed Hawks are of remarkably little interest to county officials and most representatives of the wind-power industry. The simple remedy of removing the prey base is opposed for various reasons.

Miscellaneous Raptor Perils

Raptors encounter numerous other unintended hazards. During the nesting season, nature-loving hikers, unaware of the presence of a raptor nest, may so disturb the birds (unless they are habituated to humans) that they leave their young or eggs exposed too long to the elements. Boaters may be the most significant disturbance to Bald Eagles. To protect Peregrines, certain cliffs in Yosemite Valley are closed to rock climbers during the nesting season. Northern Harriers often nest in agricultural fields, and nests and nestlings are routinely destroyed by harvesting activities.

A cluster of festive balloons, casually released miles upwind by someone oblivious of their fate, became tangled in the branches of an oak tree not more than 3 m (10 ft) from an active Red-tailed Hawk nest in the same tree. The hawks, afraid of the balloons bouncing menacingly in the wind, abandoned the nest and small young.

Bird feeders, as entertaining as they may be, are in reality an unnatural attraction that brings together a variety of species and numerous individuals, some of which are sick and infectious.

Fig. 57. A cluster of balloons, carelessly released miles away, caught up in this valley oak and caused a pair of Red-tailed Hawks to abandon their nest and newly hatched young.

Trichomoniasis, a protozoan disease commonly found in doves and pigeons, can spread to other species at feeders. The sick birds are easy prey for raptors, which in turn may become infected with the often fatal disease. Bird feeders must therefore be cleaned regularly with a bleach solution. Scores of Sharp-shinned and somewhat fewer Cooper's Hawks are killed every year by crashing into house windows situated too close to bird feeders. The effectiveness of hawk decals on windows is highly questionable. Domestic cats also are a problem.

Barbed-wire fences and overhead wires kill and injure raptors that fly into them while chasing prey; the hawks have no understanding of unyielding human-made wires (which, unlike thin branches, do not bend upon impact), or they simply may be focused on the fleeing prey and not see them. The wires strung in vineyards kill Sharp-shinned and Cooper's Hawks pursuing marauding starlings; hawks also may die of alcohol poisoning after eating birds that have fed on fermented grapes. Discarded fishing line kills raptors that become trapped in its tangles.

Electrocution, known to be a problem since the 1970s, occurs when a bird sits on a power pole and touches two wires at the same time. Raptors with long wingspans are most susceptible, and eagles, large falcons, and buteos have been killed in great numbers. Before they are released to the wild, captive California Condors are given special training to avoid power poles. Pacific Gas and Electric Company began retrofitting its poles to make them safer for raptors (by placing special perches above the wires, for example) after earning the distinction of being the first utility company in the country to be cited by the U.S. Fish and Wildlife Service for violation of the Migratory Bird Treaty Act, specifically for the deaths of Swainson's Hawks and Great Horned Owls (*Bubo virginianus*) in the San Joaquin Valley (Williams 2000). The conductivity of the new steel poles makes them even more lethal than unmodified wooden ones.

The introduced West Nile Virus, a recent (2002) arrival in California, is currently a major killer of raptors in North America.

Caring for Injured Hawks

It is a great rarity to come across a raptor that has been injured, because if it can still move at all, it is likely to hide in dense vegetation, if available; also, other predators carry off disabled hawks. Normally, a healthy hawk will not tolerate a human's close approach. If the hawk does not fly away, it is either a premature fledgling, sick or injured, or a falconer's bird.

Like other birds, young raptors, having not quite got the hang of their wings, frequently crash-land on the ground; they also sometimes fall out of their nests (or are pushed out by nest mates) when they are still quite downy and nowhere near ready to fly. If there is an obvious nest, the baby can be returned to it, except in the case of a Golden Eagle, which may be killed by its sibling if replaced. If return is impossible, the bird should be taken to one of the wildlife rescue centers scattered throughout California; only licensed rehabilitators are allowed by law to shelter raptors. Should the youngster be well feathered, with little down remaining, put it off the ground, where it is safer from predators and where it will be found and tended by its parents after your departure. Human scent left on baby birds by handling is undetectable by the adults and will not cause them to desert their young.

Sick hawks keep their plumage fluffed out even in warm weather and hang their wings and sometimes their heads, and their eyes are partly or fully closed. By contrast, recently injured raptors are alert and sleek when approached and try to run away, or they throw themselves on their backs and present their talons. An injured wing, which of course makes flight impossible, droops to the ground.

Occasionally, a raptor becomes so soaked by a rainstorm that it cannot fly, leaving it vulnerable to attacks by ground predators. A wet Turkey Vulture, caught by a dog, at first appeared dead or nearly so, its head bent under a wing (Turkey Vultures tend to go limp when handled); quite possibly it was playing possum. Placed on a safe perch, the bird, once dry, flew perfectly.

Picking up an injured raptor is frequently not easy. It is best to approach it from the rear, throw a large towel (or a blanket for very large birds) over it, and then grab it quickly while taking care to fold back the wings against the body and taking hold of the legs to preclude use of the feet. Except for the vultures, all raptors defend themselves with their sharp and sometimes highly dangerous talons (which in addition may be contaminated with bacteria), and many species bite as well.

Place the hawk in a cardboard box (small enough to keep it from thrashing around) with a towel in the bottom and plenty of air holes punched in the sides near the floor so that the hawk cannot see out; the flaps on top should be secured with tape. Never put a hawk in a wire cage, where it is likely to add to its injuries by trying to escape through the mesh. No attempt should be made to feed the hawk or to give it water; many hawks have been killed by well-meaning people feeding them cooked meat or fatty hamburger—or sometimes chicken feed. Even more appropriate food, such as mice or birds, may kill a raptor too weak to digest it. If the raptor cannot be taken to a rescue center right away, it should be kept in its box in a dark, quiet place, where it is likely to stay calm; it is kinder not to check on it periodically.

A list of wildlife care facilities can be found on the California Department of Fish and Game (CDFG) Web site at www.dfg.ca.gov/coned/guide/care.pdf, or phone that agency or the U.S. Fish and Wildlife Service for instructions on what to do with the hawk.

A hawk that allows close approach and appears healthy is likely a falconer's bird. Such hawks wear leather cuffs around their tarsi and perhaps also jesses, leg bands, or a tiny radio transmitter (the batteries of which may have failed, preventing the falconer from tracking the bird down). Occasionally, they may come down to strangers if imprinted on humans or very hungry, although most falconers' hawks can survive on their own in the wild. Falconers who have lost a hawk usually go to great lengths to find it and may spend many days searching from cars and small airplanes. Sightings of lost falconry birds should be reported to local animal control authorities, and recovered hawks wearing falconry equipment should be taken there or to a wildlife rehab center and reported to the CDFG; posting a "found" notice on the Web site of the California Hawking Club (www.calhawkingclub.org) may reunite hawk and owner. Very rarely, a raptor trailing a leash escapes; if such a bird is seen, every effort should be made to capture or at least report it, because a leash quickly becomes tangled in branches or telephone wires, a death sentence for the hawk.

Falconry

The sport of falconry probably originated in prehistoric times in the Middle East, spreading eventually to the Far East. In Europe, falconry was "the sport of kings," although not restricted to them, from the Middle Ages until the end of the eighteenth century, and princes exchanged gifts in the form of Gyrfalcons *(Falco rustico-lus)*—the costliest and most difficult to procure of all the raptors then used for hawking. In the nineteenth century, falconry was a gentleman's sport in England and persisted in very limited form there and on the continent until the middle of the twentieth century, when the art of training hawks and falcons experienced a renascence not only throughout Western Europe but also in North America. Today, the sport has a few thousand adherents in the United States, including about 600 in California, and is carried out in some ways at a higher level than ever before, thanks to the benefits bestowed by the captive breeding of raptors and the use of telemetry, which permits flying hawks at their "wild" weight without the worry of losing them.

Modern falconers, and undoubtedly those of the past, show a devotion to their charges that borders on the fanatical. Imprinted hawks, and even many that are not, eventually come to regard the falconer as a mate or a sibling; as a result, such birds are not only extremely tame but also disarmingly friendly and downright affectionate, albeit often disrespectful. Even hawks raised by their parents can become friendly and usually lack the insolence of the imprint. Add to that their spectacular aerial displays (if properly trained) and skill and cunning in capturing prey, and it is little wonder that falconers invest a prodigious amount of time in their sport. Falconry is not only the most difficult kind of hunting, but it is also highly esthetic and very spectator oriented, which is not true of gun hunting. It is important to remember that a raptor's prey has evolved over millions of years, in great part responding to predation pressure from its natural enemy, and therefore has an arsenal of escape maneuvers. The quantity of game taken by a falconer generally pales in comparison to that of a gun hunter. In addition, in falconry there is no wounded animal that crawls off to die out of sight of the hunter.

Practicing falconry is not simple. Both the U.S. Fish and Wildlife Service and the California Department of Fish and Game require a two-year apprenticeship during which the aspiring falconer is permitted to train an American Kestrel or a Red-tailed Hawk only, and only under the supervision of a master falconer. The two agencies also require an annual license, the first issued only after the applicant has passed a lengthy exam about general raptor biology and falconry. To pursue wild game, the falconer must buy an annual hunting license and appropriate stamps, such as those sold for waterfowl hunting. Lastly, and most seriously, the

Fig. 58. A successful falcon (a male Prairie-Peregrine hybrid) on his duck (a Gadwall *[Anas strepera]*), with his canine hunting partner.

falconer needs to find open areas with little human activity where the hawk can be trained, and also places to hunt, a daunting task in much of California. If the falconer wishes to go on vacation, hawks cannot simply be left in a kennel; they must be lodged with other falconers, or a knowledgeable person must house-sit or visit daily to look after the welfare of the feathered tyrant.

A commercial exploitation of trained hawks is their use in driving away birds such as starlings and gulls from airports; these birds present hazards to aviation and flee at the sight of a falcon in the air. Pigeons, too, take flight at the approach of a hawk, and Harris's Hawks have been employed in urban settings to drive them away. This use, unfortunately, suffered a setback when one of the unselective raptors seized a Chihuahua, perhaps mistaking it for a rat.

Although the CDFG holds the opinion that falconers are responsible in part for decreased populations of many of its listed species, most Species of Special Concern remain legal to take from the wild. Few of them are popular with falconers; only a handful of falconers enjoy hawking with Sharp-shinned Hawks, for example. Their nests, furthermore, are notoriously hard to find, leading officials to conclude that the species is less common than it actually is. It is true that selfish individuals in the past took young Peregrine Falcons illegally from the wild in California when that species was critically endangered (from an environmental cause); however, there is now little temptation for any falconer to engage in illegal activities, because captive breeding of nearly all raptors used in falconry has been extremely successful, and supply now exceeds demand. In fact, it was the expertise of falconers in breeding raptors that made possible the restoration of the Peregrine Falcon to most of the United States, and is currently facilitating the captive breeding of California Condors.

Recreational and Educational Activities

The Golden Gate Raptor Observatory (GGRO) records yearly fall migration counts at California's premier hawk-watching location (the Marin Headlands, just north of San Francisco), where in fall dozens of raptors may be in the sky at one time. Although the binocular count is also very high there during that season, additional volunteers are always needed. Beginners are welcome to soak up pointers from the experts, and free hawk talks and bird-banding demonstrations are offered to visiting school groups and the public. The GGRO's newsletter often features hawk-watching tips and hotspots. Its Web site (www.ggro.org) has interesting information on tracking migration progress, with daily counts and other facts, plus a raptor identification quiz. Contact the organization at GGRO, Building 201, Ft. Mason, San Francisco, CA 94123; (415) 331-0730.

Less social hawk-watchers can no doubt find very good spots elsewhere (and simply not yet discovered by others) both along the Pacific coast and inland during migration seasons. On the Web at www.hmana.org (the Hawk Migration Association of North America, HMANA), you can learn how to keep records of hawk sightings; by reporting sightings, individual hawk-watchers can add useful data to studies of migration and population trends. The HMANA Web site is also a good place to find updated lists of organized raptor-watching groups by state.

For those who like their hawks up close, volunteering at local wildlife rehabilitation centers is satisfying work. Restoring raptors to good health and releasing them generally has no impact on the populations of the various species, but the practice carries much emotional appeal and serves to familiarize the general public with birds of prey and promote their protection.

The Ventana Wilderness Society (VWS), an organization involved in the release of California Condors and Bald Eagles, offers field trips to see California Condors at Big Sur and Bald and Golden Eagles at Lake San Antonio in southern Monterey County. The society also holds Bird-a-thons and offers bird-banding demonstrations. Volunteers are trained at the VWS Big Sur Ornithology Lab. Its education program includes in-classroom presentations about migratory birds and the California Condor and bird-banding field trips. A "Condor Cam" at its rearing and release facility in Big Sur broadcasts live images of captive and wild Condors to its Web site (www.ventanaws.org); updates on the Condors are posted on the "Field Notes" Web page. Contact the society at VWS, 19045 Portola Road, Suite F1, Salinas, CA 93908; (831) 455-9514. To report a condor sighting, email info@ventanaws.org or call (831) 624-1202. For places to see Condors, try the Web site www.bigsurcalifornia.org/condors.html.

Volunteers may help track Condor movements at Hi Mountain Lookout in the Los Padres National Forest east of San Luis Obispo. The lookout may be contacted through its Web site www.condorlookout.org.

The annual Audubon Society Christmas Bird Counts offer another opportunity to demonstrate your observational talents. The counts reveal trends in wintering populations, providing data that helps government agencies make environmental decisions. Although all bird species are counted with equal devotion, volunteers who know their raptors are a valuable addition to a census team. Find your local chapter on the National Audubon Society Web site: www.audubon.org.

Some local chapters of the Audubon Society host field trips and often have some members who are particularly interested in hawks. The Kern Valley Audubon Society hosts a Turkey Vulture Festival each September. The Kern River Research Center counts the migrants through most of September and October and needs volunteers to help (www.valleywild.org/tvfest.htm).

Millerton Lake State Park (209) 822-2332 near Fresno operates boat tours to see wintering eagles, as do both the Monterey and Santa Barbara County parks departments (at Lake San Antonio (831) 755-4899, and Lake Cachuma (805) 568-2460, respectively).

Photographers and others can reserve blinds for close views of Bald Eagles at the Klamath Basin Wildlife Refuges (916) 667-2231.

The Central Valley Birding Symposium offers a raptor field

trip in the Stockton area as part of its program each November (www.cvbs.org). For up-to-date lists of birding events, some of which are raptor oriented or feature a raptor field trip, try www.birdinghotspot.com.

You can help scientists track bird movement by reporting leg-band numbers (or colored bands and tags, or colored feathers) worn by any wild species (whether found on a corpse or spotted through a scope) by phoning (800) 327-BAND. Also supply species, date, and how the information was obtained.

The Santa Cruz Predatory Bird Research Group needs volunteers to assist in an ongoing study of Peregrine nesting in California. Report new nests and the status of known sites at Falco Net@ucsc.edu. Learn more at www.scpbrg.org.

The Web site of the Santa Cruz Predatory Research Group (www2.ucsc.edu/scpbrg) allows schoolchildren and the general public to track some of California's wintering Bald Eagles, equipped with transmitters, up to their northern Canadian nesting areas in the spring and back down to California in the fall. It demonstrates how scientists can use computers and that birds seen locally have an international significance. The Web site also offers exercises and project ideas for elementary through high school students and teachers. The "Peregrine Nest Cam" video can be accessed through the site during the nesting season, and computer-generated tracking maps of individual Golden Eagles can be viewed.

Real-time video of nesting Bald Eagles, participants in the effort to reintroduce the birds to Santa Catalina Island, is broadcast over the Internet at www.iws.org from February to June. Because of the long-lasting effect of DDT off the Los Angeles coast, the eggs of these eagles are incubated at the San Francisco Zoo while the parents sit on artificial ones; the chicks are returned to the nest a week or two after hatching. Satellite tracking is also featured on this site.

Up-to-date lists of Web sites featuring nesting raptor video can be found on the Internet by using the search words "nest cam."

The Peregrine Fund Web site (www.peregrinefund.org) has good information about the reintroduction of the California Condor; its "Notes from the Field" page is entertaining and of interest.

Information about California's endangered raptors, including current threatened and endangered status, can be found at www.dfg.ca.gov. The CDFG Web site also has a page just for kids and directs educators to "Project Wild" activities for the classroom, including information on raptors.

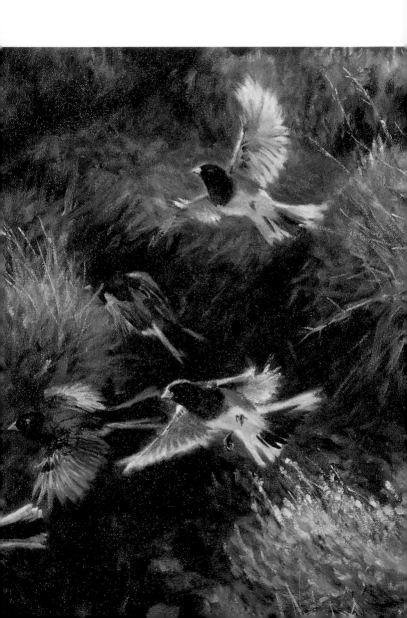

OVERVIEW AND RANGE MAP KEY

The scientific and common names used in this book conform to those used in the American Ornithologists' Union *Check-list of North American Birds,* 7th edition. In line with University of California Press policy, the official common names are capitalized: Peregrine Falcon *(Falco peregrinus),* Northern Goshawk *(Accipiter gentilis).* For brevity, sometimes these names have been shortened; whereas Peregrine remains capitalized (because it always stands for the Peregrine Falcon), goshawk reverts to lower case (it

Fig. 59. Relative sizes of (clockwise from bottom) Golden Eagle, Redtailed Hawk, American Crow, feral pigeon (Rock Dove), and American Robin *(Turdus migratorius).*

is not specific even though in California there is only one species). The sequential order also follows that of the AOU Check-list, with the exception of the Bald Eagle *(Haliaeetus leucocephalus)*.

Each species account supplies the physical and behavioral characteristics, such as shape and flight style, that are good indicators of the species. Adult and juvenal plumages are described for all species; subadult feathering is described (and illustrated in the color plates) where it aids recognition. Some major diagnostic features, in addition, are listed in the captions to the color plates, and a section on similar species in each account may help avoid confusion.

Well-marked populations of birds (and other animals) are referred to as races (or subspecies), such as the Peale's race of the Peregrine Falcon. Only races that appear in California are described. When a species or race shows variation (as of color) within a population, the variants are called morphs. An example is the rufous morph of the Red-tailed Hawk *(Buteo jamaicensis)*, which, in California, is chiefly represented by the race *calurus.*

We expect the reader to become quickly acquainted with the size of a Red-tailed Hawk and also that of an American Crow *(Corvus brachyrhynchos)*, feral pigeon (Rock Dove, *Columba livia*), and American Robin *(Turdus migratorius)* for size comparison.

The most common vocalizations, foraging, and reproductive activities are discussed where such knowledge may aid identification. More elaborate information concerning less frequently observed aspects such as the fine details of breeding and migration can be found in a number of excellent books readily available to the interested reader; generally, the most charismatic species, such as the Peregrine Falcon, have been studied most intensively, and a great deal is known about their life histories.

Seen all year
Seen in winter only
Seen in summer only
? May breed here

Overall distribution of California raptors is given here, and general seasonal movements where appropriate. In addition, where possible, we list some likely places to find each species. Range maps are provided except for the rare species, where sight-

Fig. 60. A mosaic of habitats. Seven species of diurnal raptors nest and forage in this central California landscape. Ecologic formations shown include inland cliff, oak woodland, chaparral, savanna, and grassland.

ings in California are too scattered and infrequent to make a map meaningful. Summer ranges (red) usually equal breeding ranges. Year-round ranges (purple), comprising summer and winter (blue) ranges, usually also equal breeding ranges, although birds seen in summer may be replaced by winter migrants.

Migrating individuals and wanderers may be seen in odd places, for example, Ospreys *(Pandion haliaeetus)* in Death Valley. Migration dates given are general and refer to when most migrants arrive; wanderers and early migrants may show up earlier or may remain longer.

California Habitats

For habitat designation, we generally follow the concept proposed by A. Miller (1951) of ecologic formations, which are major habitats characterized by unique or prominent vegetation or geologic features. A general acquaintance with the dominant plants is helpful, but most wildlife is attracted to ecologic formations (i.e., the type of habitat) and not to specific plants; to a meadow vole *(Microtus* sp.), for example, grassland is

Fig. 61. Coastal bluffs near Half Moon Bay (San Mateo County).

Fig. 62. Bays and mudflats attract prey species.

grassland regardless of the species of grass comprising it (excepting perhaps some unpalatable ones), and therefore it is at home in any kind of grassland as is one of its chief predators, the Northern Harrier *(Circus cyaneus).*

Many California landscapes are mosaics of habitats; for example, a pleasing vista of oak woodland often includes a riparian zone along a stream, and a cliff may tower above the trees, surrounded by a collar of chaparral. A coastal grassland may end in a precipitous line of cliffs to the west and stop at a wall of dense coast coniferous forest inland. Such diversity provides for

Fig. 63. Coastal grassland.

a wide range of potential prey animals, which in turn attract a variety of hawks. Because not all ecologic formations are equally attractive to hawks, only the most important ones for the purposes of this book are listed and briefly described below, each with its raptor community. Others are arbitrarily grouped together.

COASTAL BLUFFS, CLIFFS, AND BAYS: Long stretches of the California coast drop abruptly to the sea in ramparts of rocky cliffs, as at Big Sur (Monterey County), or with steep bluffs and headlands as at Año Nuevo State Park (San Mateo County). There are also solitary prominent rocks, such as Morro Rock in San Luis Obispo County (site of the best-known Peregrine eyrie in the state). Rocks and bluffs are generally poor foraging areas but are much favored by various raptors in migration and are a major haunt of Peregrine Falcons, which like to launch themselves at quarry from a height. Large bays that are bordered by mudflats at low tide are also sought out by Peregrines and Merlins *(Falco columbarius),* both species drawn by the enormous numbers of waders that utilize these "cafeterias" during migration and the winter months. The mouths of northwestern rivers also draw these falcons, along with numerous Bald Eagles and Osprey.

COASTAL GRASSLANDS: Grasslands occur in places along the central and northern California coast; much of Point Reyes (Marin

Fig. 64. Coastal salt marsh. A high winter tide has created a rich feeding ground for some raptors as rodents are forced to the tops of the plants.

County) and Salt Point (Sonoma County) are representative of coastal grasslands (prairies), which stay greener than interior grasslands. Numerous small rodents and birds attract White-tailed Kites *(Elanus leucurus)*, wintering buteos, American Kestrels *(Falco sparverius)*, and Merlins; Ospreys commute overhead to and from fishing areas, Northern Harriers breed and winter here, and even Golden Eagles *(Aquila chrysaetos)* are seen on occasion. Updrafts provided by the onshore winds make this habitat highly attractive to various raptors.

COASTAL SALT MARSHES: These marshes are periodically inundated by tides, and the salt content of the water greatly limits the variety of plants that survive here, with the most obvious being pickleweed (*Salicornia* spp.). Although much reduced by development, some salt marshes persist around and near San Francisco Bay, Humboldt Bay, and Morro Bay. Salt marshes have tidal channels and pools and, often, islands of grass and grassy levees, all of which provide habitats for a variety of rodents and water birds that in turn attract Northern Harriers, some White-tailed Kites, buteos, Merlins, and Peregrine Falcons.

COASTAL SCRUB: A discontinuous band of scrubland covers some of the coastal hills and headlands, with coyote bush *(Baccharis pilu-*

Fig. 65. Coastal scrub, coastal grassland, and coastal coniferous forest.

laris) dominating along the north and sage in the south. Judging by the numbers of buteos kiting over the bushes, there is clearly an abundance of small rodents and rabbits here that are sufficiently accessible despite the dense vegetation, making this habitat a magnet for wintering hawks. Sharp-shinned Hawks *(Accipiter striatus)* and Cooper's Hawks *(A. cooperii)* are also attracted by the profusion of migrant sparrows from August until April. Kites and harriers are common. Coastal scrub is conspicuous along the coasts of San Mateo County and along long stretches of the southern California coast, from Santa Barbara southward.

COAST CONIFEROUS FOREST: Found along the northern two-thirds of the California coastline are two major types of coastal forests, one dominated by redwood *(Sequoia sempervirens)* and Douglas-fir *(Pseudotsuga menziesii),* as in Humboldt County, and the other by several species of pines that grow in often highly local populations, as on Point Reyes and near Monterey. These generally moist forests, where you would expect abundant prey, do not have a great variety of raptor species; Sharp-shinned Hawks and Cooper's Hawks are found here, along with American Kestrels along the edges, and Turkey Vultures *(Cathartes aura).* Where rivers cut canyons through the forests, Ospreys, and sometimes Bald Eagles, are conspicuous.

FRESHWATER MARSHES: Rivers and creeks often spread into deltas before entering the Pacific Ocean, and here cattails and tules

Fig. 66. North coast river bordered by slopes of alder and coniferous forest.

form marshes. The Central Valley has remnants of enormous marshes that provide habitat and food for many thousands of wintering waterfowl as, for example, near Los Banos (Merced County) and Gray Lodge Refuge (Butte County). There are also large marshes on the Modoc Plateau. Open areas of water make marshes especially attractive to ducks, which may also nest here. Not surprisingly, harriers and Peregrine Falcons are attracted, as well as some White-tailed Kites.

RIVERS AND LAKES: The northern mountainous parts of California are laced with numerous rivers, and lakes and reservoirs are found the length of the state. Although some of the northern rivers hold sufficient fish to support nesting populations of Ospreys and Bald Eagles in spring and summer, in winter these species join Golden Eagles, Peregrine Falcons, and Red-tailed Hawks at lakes and reservoirs through much of the state. Peregrines also nest on cliffs above lakes and rivers.

INTERIOR GRASSLANDS AND SAVANNA: Much of California's grassland has been converted to farmland, but large grassy tracts persist in lake basins in northeastern California, in both the Sacramento and San Joaquin valleys, and on the Carrizo Plain, and smaller areas are scattered through much of the state. Savannas (see fig. 1) are grasslands with widely scattered trees (which provide nest sites for raptors)—commonly valley oaks *(Quercus lo-*

Fig. 67. Chaparral, here composed mainly of manzanita and chamise, with scattered gray pine and blue oak.

bata), although human-made savanna fragments with nonnative trees are often found around corrals and abandoned homesteads. Both grasslands and savannas support a great variety of raptors, including Turkey Vultures, California Condors *(Gymnogyps californianus)*, Northern Harriers, White-tailed Kites, surprising numbers of small accipiters, Red-tailed Hawks and Swainson's Hawks *(B. swainsoni)*, wintering Ferruginous Hawks *(B. regalis)* and Rough-legged Hawks *(B. lagopus)*, Golden Eagles, American Kestrels, and Prairie Falcons *(Falco mexicanus)*.

CHAPARRAL: Thousands of acres of California hillsides and mountain slopes are covered by chaparral, dense brush that is often a mosaic of various kinds of bushes and minitrees. In the Sierra foothills, manzanita *(Arctostaphylos* spp.) may form the dominant species; in the Coast Ranges, coyote bush, deerbrush *(Ceanothus* spp.), or chamise *(Adenostoma fasciculatum)* are examples of dominant plants. Potential prey includes rabbits and various rodents such as mice and pack rats, along with birds and reptiles. Turkey Vultures, Redtails, Golden Eagles, Sharp-shinned and Cooper's Hawks, and Prairie Falcons forage in chaparral.

OAK WOODLAND: Ringing the Central Valley, oak woodland covers much of the surrounding foothills as well as much of the Coast Ranges, especially on north-facing slopes. It is conspicuous east

Fig. 68. Oak woodland is frequently interspersed with other ecologic formations, such as grassland.

of Sacramento for example, in parts of the San Francisco Bay Area, and also in the hills near San Diego. Different oak species dominate in different areas, depending on elevation; often, a scattering of other trees are mixed in, such as California bay *(Umbellularia californica)* with coast live oak *(Quercus agrifolia),* or gray pine *(Pinus sabiniana)* with blue oak *(Q. douglasii).* Such woodland has a fairly open canopy easily penetrated by the sun in most places so that the soil dries out in summer yet permits grass to grow in spring. Frequently, oak woodland is interspersed with meadows and chaparral. Turkey Vultures, California Condors, and Sharp-shinned, Cooper's, and Red-tailed Hawks all can be found here, as well as Golden Eagles and American Kestrels.

RIPARIAN WOODLAND: Ribbons of trees that edge permanent or ephemeral bodies of water form riparian woodland, which are found from the mountains to the deserts. The trees can be fairly well spaced, as are the western sycamores *(Platanus racemosa)* that grow along gravelly arroyos, or they can be packed together, like willows *(Salix* spp.). Alders, both red and white *(Alnus rubra, A. rhombifolia),* cottonwoods *(Populus* spp.), and even fan palms *(Washingtonia filifera)* and the palo verde *(Cercidium* spp.) can be components of riparian woodland. Although this habitat is obviously limited in extent, it nevertheless is home to a

Fig. 69. Riparian woodland, here composed chiefly of red alder, western sycamore, and willows.

great variety of birds and other animals that are preyed on by raptors, and it may be the nearest nesting habitat for some hawks that are actually birds of the grassland and other habitats. Riparian woodland is popular with Turkey Vultures, White-tailed Kites, Cooper's Hawks, Swainson's Hawks, Red-tailed Hawks, Red-shouldered Hawks *(Buteo lineatus)*, Harris's Hawks *(Parabuteo unicinctus)*, and American Kestrels.

MONTANE AND SUBALPINE FORESTS: California's mountains are covered, up to timberline, with forests of chiefly coniferous trees that change in density and to some extent in height depending on elevation. Lower down, the first stands of conifers make up montane forest, distinguished by its relatively open parklike aspect, the well-spaced trees chiefly ponderosa pine *(Pinus ponderosa),* Jeffrey pine *(P. jeffreyi),* and incense-cedar *(Calocedrus decurrens);* in higher montane forest, white fir *(Abies concolor)* takes over as the dominant tree. This forest is frequently intermixed with chaparral. Good examples are seen in the Siskiyou and Trinity Mountains, the lower Sierra Nevada, and the San Bernardino Mountains. Rodents such as various chipmunk species and squirrels, along with a variety of birds, provide prey for raptors.

At higher elevations, subalpine forest replaces montane forest; it is generally much denser, and the dominant trees are red fir *(Abies magnifica)* and lodgepole pine *(Pinus contorta),* the latter forming thick stands. Because of abundant snow, this habitat is well watered later in summer, with many wet meadows, lakes, and streams seamed with quaking aspens. Examples of subalpine forests can be seen near Donner Lake (Nevada County), at the higher elevations of Mt. San Jacinto (Riverside County), and in the San Bernardino Mountains. Marmots *(Marmota flaviventris)* and other rodents, Snowshoe Hares *(Lepus americanus),* and birds such as woodpeckers, jays, and grouse are typical raptor prey. Food becomes scarce in winter in subalpine forests, less so in montane forests.

Forest meadows attract typical grassland raptors such as American Kestrels, Red-tailed Hawks, and Northern Harriers. Other raptors of both of these mountain forests include Turkey Vultures, Northern Goshawks, Sharp-shinned and Cooper's Hawks, and Golden Eagles.

SAGEBRUSH (SAGE STEPPE) AND PIÑON-JUNIPER WOODLAND: The eastern side of the Sierra Nevada drops steeply and forms one of the walls of the Great Basin, a bowl that lies in the rain shadow of the mountain range and therefore receives little precipitation. The vegetation is drought adapted and not very diverse, the most abundant and obvious plants being big sagebrush *(Artemesia tridentata)* and, to a lesser extent, rabbitbrush *(Chrysothamnus* spp.). Although individual bushes are not as tightly packed as in

Fig. 70. Subalpine forest, near timberline.

chaparral (making prey animals more visible in the alleyways between them), each bush provides ready cover for small and medium-sized birds and mammals.

Sagebrush is in places interwoven with piñon-juniper woodland, a very open form of woodland composed of piñon pine *(Pinus quadrifolia),* western juniper *(Juniperus occidentalis),* or both and, in some areas, Joshua trees *(Yucca brevifolia);* these serve as nest sites for some of the raptors that hunt here and over sage. Sagebrush is seen from Inyo County northward, and examples of piñon-juniper woodland are found on top of some of the mountains of the Mojave Desert, in the White Mountains, and extensively on the Modoc Plateau. Raptors seen here include the Turkey Vulture, Northern Harrier, Red-tailed and Swainson's Hawks, Golden Eagle, American Kestrel, and Prairie Falcon. Accipiters are encountered at times as well, particularly during migration, and Northern Goshawks forage extensively in sagebrush on migration in fall and when they nest in the riparian woodland or quaking aspens commonly found on the eastern Sierra slope. Ferruginous and Rough-legged Hawks may winter in this habitat.

DESERTS: Two rather distinct deserts, the Mojave and the Colorado, take up most of southeastern California; a much reduced area of desertlike habitat is found in the southern San Joaquin Valley, where saltbush *(Atriplex* spp.) provides cover for small animals such as reptiles and rodents. The Mojave is dominated by a single plant, the

Fig. 71. Sagebrush and piñon-juniper woodland.

Fig. 72. Colorado Desert with creosote bushes, here intermingled with paler burro-weed *(Ambrosia dumosa).*

creosote bush *(Larrea tridentata),* which covers hundreds of square miles from which arise stark mountains. The warmer Colorado Desert has a greater variety of shrubs, including creosote bush but also small trees such as blue palo verde *(Cercidium floridum)* and ironwood *(Olneya tesota).* Wildlife is not abundant except for some insects, reptiles, a few birds and small mammals, the latter, however, chiefly nocturnal and therefore inaccessible to hawks. Raptors are therefore not numerous; they include the Turkey Vulture, Swainson's and Red-tailed Hawks, Golden Eagle, American Kestrel, and Prairie Falcon. Golden Eagles nest in the Granite Mountains (San Bernardino County) and also, along with American Kestrels and Prairie Falcons, in the northern part of Anza-Borrego State Park (San Diego County), where all use cliffs.

INLAND CLIFFS: Sprinkled throughout California are cliffs and rock outcrops large and small, famous cliffs like El Capitán in Yosemite National Park (and another El Capitán near San Diego), great towers of rock such as in Pinnacles National Monument in San Benito County, and clifflets tucked away in the canyons of the Coast Ranges. All of them provide the potholes, niches, and ledges used as nest sites by an elite group of California raptors: the California Condor, the Golden Eagle, Peregrine and Prairie Falcons, as well as the Turkey Vulture and American Kestrel; for Turkey Vultures, holes between and under fallen boulders at the base will do.

Fig. 73. Inland cliff. Peregrine Falcons have used one of the potholes for their eyrie.

Fig. 74. A reservoir surrounded by interior grassland, savanna, and oak woodland. New reservoirs have allowed some raptors to increase their ranges.

HUMAN-MADE HABITATS: An astonishing variety of environments made by humans have been adopted by raptors because of the abundance of prey that can be found there. Highway shoulders and center strips are very attractive to Red-tailed Hawks, White-tailed Kites, and American Kestrels. Alfalfa fields are usually teeming with small rodents that are food for these birds as well as for Rough-legged and Ferruginous Hawks and Northern Harriers.

The ponds of duck clubs and urban lakes such as Oakland's Lake Merritt (Alameda County) and lakes in the state's major zoos attract not only waterfowl but also numbers of pigeons and small birds so that Sharp-shinned Hawks, Cooper's Hawks, and Peregrine Falcons are often found here or nearby. Reservoirs are attractive to both Golden and Bald Eagles, both species nesting near them and dining on coots and other waterfowl in winter.

Metropolitan garbage dumps with large gull and starling populations also attract Peregrine Falcons, along with Merlins. Merlins are very fond, too, of the crowds of sparrows, blackbirds, and starlings that hang around dairy farms operating, for example, in parts of the Central Valley and in Sonoma County. Windbreaks around farms provide perches and nest sites, the most common being rows of eucalyptus that seem to always hold a Red-tailed Hawk nest along with that of an American Kestrel, a species that puts the clumps of leaf and bark debris lodged in forks to good use. Golf courses and airports, pockets of remaining open space in urban environments, are also attractive to raptors. At the San Jose Airport (Santa Clara County), a Gyrfalcon *(Falco rusticolus)* stooped several times at an American Kestrel that had become entangled by its feet, only to be harassed in turn by a Peregrine that was subsequently struck and killed by a commercial jet; all the while, a Prairie Falcon circled overhead (Balgooyen 1988).

With the sharp decline of hawk shooting in the late 1960s, many raptors have become remarkably unconcerned about the close proximity of humans, and urban areas have become veritable meccas for birds of prey, offering new and nearly unexploited foraging and nesting habitats. For a Peregrine Falcon, a high-rise or soaring bridge is an acceptable artificial cliff, especially if there is an endless supply of city pigeons nearby, American Kestrels quietly slip into their nest cavities in the skirts of dead fronds hanging below the crowns of stately fan palms, and in Hayward (Alameda County), a pair of Red-tailed Hawks raises young every year in a minipark at the intersection of two major streets. Wintering Sharp-shinned Hawks and Merlins take advantage of the bounty of small birds that assemble at bird feeders. Turkey Vultures do not seem to mind noisy traffic and barbecues and roost in street-side and backyard trees, often decorating roofs, sidewalks, and patios with white splashes, molted feathers, and wads of regurgitated fur.

Fig. 75. Turkey Vultures sunning.

Vultures

Nearly all raptors eat carrion at some time or another, but only those that do so exclusively (or nearly so) have conspicuous adaptations for that diet. They are resistant to many toxins and pathogens. Some, including the Turkey Vulture *(Cathartes aura),* have a well-developed sense of smell, which is rare among birds; it helps them locate hidden carrion. Fresh food is, however, preferred.

Like most of their Old World counterparts, California's vultures have nearly bare heads, presumably to facilitate cleaning after feeding in the body cavities of decomposing carcasses; oddly, their heads are also adorned with numerous deep folds and creases that would seem ideal pockets for growing bacteria. Because of its lack of feathers, the head of a flying vulture appears very small—so small, in fact, that at a distance, the bird appears headless; this simple visual clue establishes at once from afar that the raptor in question is not an eagle, with which vultures are frequently confused. The appearance of headlessness is enhanced by their habit of flying with the head hanging down at an angle, a characteristic apparently common to all New World vultures.

Fig. 76. A young Turkey Vulture in an abandoned mine shaft, alarmed by an intruder, regurgitates food. Adults may also do so.

A perched vulture can draw its head back into the skin of its nape so that a crescent of feathers bristles upward, giving the bird the appearance of a cabaret singer luxuriating in her feather boa; it may keep the neck and head warm and may also possibly be for display. To conserve energy, Turkey Vultures drop their body temperature at night and, upon awakening, shiver and quickly regain their day temperature (Heath 1962). Perched vultures are also in the habit of fully spreading their wings, especially in early morning, either facing the early sun or away from it; hypotheses for the function of this behavior include thermoregulation, drying off, plumage disinfection, detection of slight air currents to facilitate soaring, and warming up, the last being the most likely. New World vultures form one of the bird families that practice urohydrosis, squirting excreta onto the legs during hot weather to promote evaporative cooling.

When alarmed, vultures throw up recent meals, perhaps for weight reduction to make takeoff easier, but in many cases vomiting appears to be an emotional reaction. Disturbed nest-bound young, too, regurgitate often putrid food, with the coincidental effect of deterring would-be predators.

BLACK VULTURE
Coragyps atratus

Pls. 1, 10   `RARE`

IDENTIFICATION: Large body, all black except for the outer five or six primaries, which are dirty white rimmed with black. The tips of the folded wings extend to the tip of the tail or beyond. The legs are pale gray. The small head is gray black, naked, and strongly wrinkled in the adult; the beak is dark with a whitish tip. The juvenile is like the adult, but the head and neck have little or no wrinkling and some blackish down; the juvenile's beak is all dark.

FLIGHT: Active flight is a series of hasty, choppy wingbeats followed by gliding. A Black Vulture flies steadily, without teetering and soars with almost flat wings, often at great heights. The long, broad wings have a conspicuous white patch on the hand portion. This raptor wing-flexes (see fig. 77). The tail is square but may look rounded, with pointed corners when fully fanned.

VOICE: The Black Vulture is a normally silent species that may hiss or grunt, usually in disputes over food.

FEEDING: Lacking the olfactory talents of the Turkey Vulture *(Cathartes aura),* the Black Vulture relies on vision to locate food and watches the behavior of other vultures in their quest for carrion. It forages mainly on the wing, often much higher than the Turkey Vulture. It also kills small mammals such as opossums and skunks, as well as livestock, especially newborn young and females debilitated from giving birth. A large flock killed a tethered pitbull (J.-A. Otero, pers. comm., 2004). It is quarrelsome and aggressive at a carcass.

REPRODUCTION: Courtship and nest sites resemble those of the Turkey Vulture. The Black Vulture does not breed in California.

DISTRIBUTION AND HABITAT: The Black Vulture is widespread in the southeastern United States, Mexico, and Central and South America. Southern Arizona is also part of its normal range.

The very few Black Vultures seen in California are, apparently, visitors from Mexico. They show no obvious habitat preference; records are from northern (Humboldt and Butte Counties) and southern parts of the state.

SIMILAR SPECIES: Although resembling a Turkey Vulture, the Black Vulture is slightly smaller though heavier. The Turkey Vulture has a longer tail and in flight lacks white hand patches and typically flies with its wings in a dihedral.

TURKEY VULTURE   *Cathartes aura*
Pls. 1, 10; Figs. 2, 7, 55, 75–78

IDENTIFICATION: Very large, bigger than a Red-tailed Hawk but smaller than an eagle, with an elongated body, a tiny-looking head, and a moderately long, rounded tail. The tips of the folded wings reach the tip of the tail or beyond. The long, rose-colored legs are usually whitened by excreta and are close together, giving the bird a somewhat knock-kneed appearance. An adult Turkey Vulture at close range appears disheveled and unkempt, in contrast to the sleeker or softer lines of true hawks and eagles; the upperwing coverts frequently appear somewhat out of alignment—probably because of the protracted molt, which causes gaps in the feather rows.

The adult's naked head is red and wrinkled, sometimes with a few white warts in front of the eye. The beak is ivory white (dark tipped in the second year). The back and folded wings are black brown; the feathers have tan margins and a slight purple gloss. The underparts are black.

The juvenile is like the adult, but the head is unwrinkled and gray black with some dark down. The beak is dark, the feather

Fig. 77. Turkey Vulture, showing wing flex. Probably all raptors wing flex to some degree when gliding into a stiff breeze.

gloss is much reduced, and the narrowly buff-edged upperwing coverts appear neatly aligned.

FLIGHT: The Turkey Vulture looks headless from a distance. Typically, the wings are held in an often-pronounced dihedral when gliding, flex gliding, or soaring on breezy days, teetering from side to side. On calm days a Turkey Vulture often waits for the sun to warm the air and provide better lift. If it takes to the wing in the absence of sunlight (and especially on damp, still mornings), the bird teeters only rarely; the wings are then held horizontal or very nearly so, and the tail is fully fanned, which it normally is not. On windy mornings, these vultures may fly well before sunrise.

This species frequently flexes its long wings downward while gliding. In a glide, the tail looks longish, but it appears short when the bird is soaring. When buffeted by a wind, a Turkey Vulture appears very lightweight, the body seemingly hanging from the wings; it lacks the heft of large hawks and eagles.

The silvery undersides of the flight feathers may appear bronze or golden because of reflections from the ground. The tail is not as silvery as the remiges.

VOICE: Although generally silent, a Turkey Vulture may at times hiss or produce growling sounds.

FEEDING: A Turkey Vulture forages on the wing, usually flying fairly close to the ground. It is adept at flying through woodland and is aided in its food search by its acute sense of smell; it has been seen to dig up buried carcasses to eat them (Smith et al. 2002). This vulture is quite happy making do with small items such as mice, road-killed snakes, and even insects; its low wing-loading is a great energy saver and makes coming to ground for such tidbits worthwhile. Dead fish make a meal, and ground squirrels and snakes killed by cars are very common food items in California. There are some observations of the Turkey Vulture killing and eating incapacitated or young birds and even young pigs. It has been seen to eat pumpkins, and other plant material turns up in their regurgitated pellets; it sometimes eats cow dung and other fecal matter, presumably for their maggot and beetle content and proteins from shed intestinal cells.

REPRODUCTION: Courtship displays involve pair flying, the presumed male closely following its mate, banking and turning together, often in descending, curving flights. Spread-wing displays before or after copulation may be performed by both sexes. No nests are built; the eggs are laid on the floor of caves, abandoned

Fig. 78. Cave used by Turkey Vultures as a nest site.

mines and sheds, niches in banks and cliffs, against the base of or under boulders, in tree cavities and hollow logs, and in thickets.

Two eggs are the usual clutch size, incubated for about five weeks. The young fledge about nine weeks after hatching.

DISTRIBUTION AND HABITAT: The range of the Turkey Vulture extends from southern Canada and across most of the United States to the southern tip of South America. Most North American birds spend the winter in South America. In the United States, year-round populations occur in the southeastern states, in parts of southern Texas and Arizona, and in western and southern parts of California.

Except in winter, the Turkey Vulture is ubiquitous in California and can be seen foraging over all terrestrial habitats; it nests throughout the state.

Northern birds join California birds as early as late July, and northbound vultures appear in southern California as early as mid-January. A great many spend the winter from central coastal California southward.

SIMILAR SPECIES: The Golden Eagle *(Aquila chrysaetos)* and Bald Eagle *(Haliaeetus leucocephalus)*, from a distance, have obvious heads and do not teeter in flight. The Zone-tailed Hawk *(Buteo albonotatus)*, which is very rare, has a big head, yellow cere and gape, and bars in the tail. A California Condor *(Gymnogyps cali-*

fornianus) is much bigger, with mottled or white underwing coverts.

REMARKS: The Turkey Vulture soon becomes familiar to the hawk-watcher, and because of its abundance, it often serves for comparison with other raptors. Turkey Vultures are interesting to watch at their roosts, being given to hierarchical interactions among themselves and to leisurely preening, sunning, and stretching before getting under way for the day. This species looks heraldic when it spreads its wings to the sun, and the hissing and rustling sounds of its pinions as it approaches a roost close overhead attract attention.

A Turkey Vulture is rarely seen alone; although other raptor species congregate at times, no other raptor does so year-round, except perhaps with its mate. Even Turkey Vulture courtship flights may be done in the company of several other individuals. Night roosts may include hundreds of individuals; in recent years, such gatherings have appeared in California towns and are not always welcome.

The scientific name is rather interesting. "Cathartes" is taken from the Greek *kathartes,* meaning "cleanser"; "aura" sounds temptingly like the Latin *aureum,* meaning gold, but is most likely derived from the Carib aboriginal name for the species, which is known even today in Cuba as *el aura.*

CALIFORNIA CONDOR *Gymnogyps californianus*
(Pls. 1, 10; Fig. 80) RARE

IDENTIFICATION: Huge and generally black. Except for a patch of very short, small, black feathers on the forehead, the adult's otherwise naked head is pink and yellow in color; air sacs in the throat and cheeks and in the blue and pink neck are often inflated. A ruff of long, black lanceolate feathers at the base of the neck is erectile and can nearly enclose both head and neck. The back and folded wings are mostly black, with the coverts adjacent to the secondaries in part white or gray, forming a short, broken bar on the wing. Secondaries below the bar are gray. Underparts are mostly slaty to black, but axillaries are white, continuing into white bars that taper toward the wingtip, formed by underwing coverts.

The juvenile is like the adult, but the head is gray black, the head and neck appear noninflatable, the wing bars are faint, and the underwing covert markings are mottled, not clean white. The feathering and head colors become more adultlike as the bird ages.

FLIGHT: The California Condor has an immense wingspread and a short tail. In a soar, the primaries show long "fingers." The wings are usually held slightly above the horizontal while soaring; in windy conditions, they may be arched forward and curved down somewhat, with the hands cupped forward a bit, a position that prominently displays the spread fingers. Compared to other raptors, a Condor can soar at a glacially slow pace. A Condor flies steadily like an aircraft, occasionally flexing its wings while soaring or gliding. This wing flex has been called a "double dip" for reasons that are not clear.

VOICE: The California Condor appears to lack calls and expresses itself only through occasional grunts, belches, and hisses.

FEEDING: A Condor forages on the wing, generally under 150 m (500 ft) in altitude, and may fly 50 km (30 mi) or more without landing, at times at altitudes of 600 to 900 m (2,000 to 3,000 ft) or more, though over peaks it rarely rises above 150 m (500 ft). It takes full advantage of winds, updrafts, and thermals, which perhaps is why it is generally a late riser and sets out later than most raptors.

Although it may feed on small dead animals such as ground squirrels, the Condor prefers large carcasses, often locating them by sight or by following Turkey Vultures *(Cathartes aura)*. In groups, they may displace a carrion-feeding Golden Eagle *(Aquila chrysaetos)*, but more often the Golden Eagle is dominant, with a single individual able to stand off multiple Condors. When California Condors were more numerous, feeding aggregations of two dozen or more were seen at a single carcass and, having fed, would wait for their companions to finish before heading back to the roost together. Some, however, seem to have strong dislikes of certain individuals and chase them away from a carcass or even engage in knockdown tumbling and biting brawls (Koford 1953).

While pulling on carcasses, a Condor may lower itself on its heels for better leverage. The serrated edges of its tongue facilitate stripping meat from bones. The California Condor is not known to attack live animals, although in the nineteenth century some were seen to eat mussels. Young Condors have been observed eating greenery, including grasses and manzanita *(Arctostaphylos* spp.) leaves, and they also sometimes swallow bits of shell, presumably to obtain calcium (Snyder and Schmitt 2002). It is thought that ingestion of bottle caps and plastic bits derives from shell-eating; the Black Vulture *(Coragyps atratus)* also eats foreign objects. Typically, however, carrion is the standard diet, including animals as large as whales.

REPRODUCTION: Courtship displays include pair flying, where both partners glide close to each other and turn together, and a rather solemn display on the ground. The presumed male spreads his wings perpendicular to the body, with primaries hanging, droops his head, and slowly turns from side to side or in a circle; he may hobble toward the female, turn, and shamble away. The female is said to act unimpressed.

Both partners select a nest site, usually a pothole or cave in a cliff; sometimes a burned-out hole in a large tree is used. No nest is built, although pebbles and debris on the cave floor may be moved about.

Condors reach sexual maturity at seven years and then commonly produce only one young every other year; the egg is incubated for about eight weeks, and the young fledges five months after hatching. Even then, the young depends entirely on its parents for food for another two months or so and is then semidependent for about four months more (Snyder and Hamber 1985).

DISTRIBUTION AND HABITAT: California Condors have been released in the Sespe Wilderness (Ventura County) and in the Los Padres National Forest. Birds released near Big Sur (Monterey County) often fly south to join others at Sespe and vice versa to a lesser extent (J. Burnett, pers. comm. 2003). Pinnacles National Monument (San Benito County) is a new (2003) release site in California; Condors also have been introduced in Arizona and reintroduced in Baja California. California's wild population (in 2004) numbered fewer than 50 birds; approximately 70 more are in captivity within the state, with captive young being raised successfully every year. More than 125 others, captive and wild, reside outside of California.

Locating a Condor is not easy. Pinnacles National Monument and the Big Sur coast offer the greatest potential of seeing one of these very rare birds. With a bit of luck, they can be seen above the private campground near the monument's east entrance, especially in late afternoon. Near Big Sur, they can occasionally be spotted from Hwy. 1 and are sometimes seen on winter mornings roosting in tall trees near the Pfeiffer–Big Sur State Park nature center; during the day, these birds fly along the hills to Julia Pfeiffer Burns State Park (look along Ewoldsen Trail). A bit farther south, birds may be found at Cone Peak in Los Padres National Forest and the Sespe Wilderness and environs such as Piru Creek and perhaps the McGill Observation Point on Mt. Pinos, both in Ventura County. The Mt. Pinos observation point is a traditional

Fig. 79. A California Condor's primary, worn here as a prop by a ranger during a Condor release at Pinnacles National Monument, hints at the immense size of this species.

site for watching Condors, especially from August through October, when the birds presumably seek out the cool temperatures of higher elevations (Snyder and Snyder 2000).

Rocky canyons and shorelines and expansive ranchlands of grass and savanna are typical habitat.

SIMILAR SPECIES: A Turkey Vulture is much smaller, lacks mottling or white on underwing coverts, has a longer tail, and has unsteady flight. A Golden Eagle is smaller, has a conspicuous head and pale nape, and is brown.

STATUS: Listed as endangered both federally and in California.

REMARKS: A California Condor can have a wingspan of nearly 3 m (about 9.5 ft) and averages about 8 kg (about 19 lb); a Golden Eagle by comparison can have a wingspan of a little over 2 m (about 7 ft) and weighs 4.5 kg (10 lb) on average. When side by side, as at carrion, a Condor dwarfs a Golden Eagle.

Fig. 80. California Condor soaring, with wing tags prominent. Photo by R. Kerr.

The Condor bathes frequently; afterward, and also on cool mornings, it spreads its wings fully as it suns itself. When too hot, it cools down by voiding on its legs, panting, and inflating the air sacs of the head and neck. Young Condors like to hang out at water holes to socialize in groups, often bathing together.

Perhaps the easiest way to identify a Condor is to look for the large, colored wing tags bearing a number (and small radio transmitter); all individuals are so equipped. Some of the most recently released birds sport global positioning system (GPS) transceivers and are tracked by satellite. Numbered leg bands, which are routinely affixed to released raptors, are unsuitable for vultures because voided urinary wastes (urates) build up inside the band, eventually constricting the blood flow.

Osprey

The Osprey is so highly specialized that some ornithologists place it in its own family (Pandionidae); more often it is lumped with the hawks and eagles. It is found on all continents except Antarctica, as well as on numerous islands.

OSPREY *Pandion haliaeetus*
(Pl. 2; Fig. 81)
IDENTIFICATION: Nearly the size of an eagle. In the adult, the head

is white, but the white crown has black brown median streaks, and a broad, blackish stripe runs through the eye. The back of the head has a shaggy crest. The back and folded wings are dark brown. The white underparts have a necklace of spots or streaks curving across the breast (always in the female but only sometimes in the male; the sexes are alike otherwise, the male being slightly smaller). An Osprey has a striking underwing pattern, the coverts white, the gray-barred flight feathers bordered by black brown coverts that also form a conspicuous large patch on the underhand. The cere and long legs are blue gray.

The juvenile is similar to the adult, but the feathers of the back and folded wings have conspicuous whitish edges so that the hawk's upper surface looks scaled.

An Osprey usually perches in a horizontal or diagonal position; it has a hunched appearance even when erect. The wingtips reach beyond the tip of the tail.

FLIGHT: The Osprey glides and flies with wings angled like a gull's, wrists projecting and hands curving back and down. It soars with wings held in a flat M shape (when seen head-on). In flight, the

Fig. 81. Ospreys often nest on human-made structures. A fledgling exercises its wings while its parent (right) tries to stay out of the way.

Fig. 82. The nests of Ospreys are usually conspicuous, regardless of location (here on the Russian River).

very long but rather narrow wings have a distinctive bend at the wrists.

VOICE: Calls include plain whistles and chattering "chip" calls given in alarm.

FEEDING: The Osprey feeds nearly exclusively on fish and hunts chiefly by gliding and soaring at no great height over water, often hovering before stooping. Occasionally, an Osprey dives from perches. The hawk drops headfirst and then throws its feet forward so they enter the water before the head and body in a sort of stiff-arm attack (see pl. 2). An Osprey tries for fish down to depths of about 1 m (3 ft). It rises with its quarry from the water surface with a powerful wingbeat, orients the fish in its talons so that it faces forward, and shakes the water from its feathers. The fish is carried to a snag or other perch and eaten.

Although the Osprey prefers fish in the 200- to 400-g (7- to 14-oz) range, it also feeds on frogs, turtles, and small mammals; birds form a negligible part of the diet. One biologist, using a tethered pigeon to trap migrant Peregrines for banding, was startled by an Osprey as his first customer, attracted no doubt by the lure bird's struggling.

REPRODUCTION: The exuberant courtship displays include circling together while calling, and undulating flights by the male. Nest sites are most often very near water and can be trees with dead tops, telephone poles, high-tension towers and other human-made structures, and even the ground on islets where water protects the site. Ospreys may nest in loose colonies. Although the sometimes huge nest of this hawk is largely a structure of sticks, the species is famous for including all manner of odd building material, such as bits of plastic tarps, old flags, and Styrofoam cups.

The clutch size is usually three eggs, which are incubated a little over five weeks. The young fledge at about seven and a half weeks of age.

DISTRIBUTION AND HABITAT: This species is nearly cosmopolitan, if not as a breeding bird then certainly in winter.

The Osprey is seen throughout California (almost always by water) in places where there is an adequate prey base, but the highest populations (year-round) are found in the northern half, particularly along the rivers of the Coast Ranges from Marin to Del Norte Counties and in the northern mountains. Those in southern California are mostly winter visitors or migrants chiefly passing through from November to April and the majority are seen along the coast. In September, numerous young birds, apparently heading southward, are seen in the Los Angeles area.

Typical habitat includes lakes and reservoirs, bays and

seashore, and rivers; nonnesting individuals may appear on surprisingly small creeks.

Important nesting areas include Lake Shasta (Shasta County), Eagle Lake (Lassen County), and Lake Almanor (Plumas County). Ospreys are also seen near the Baum Lake Fish Hatchery (Shasta County); along the Russian River (Sonoma County) and Eel River (Mendocino County); and at Tamales Bay, Nicasio Reservoir, and Kent Lake (Marin County). Large concentrations of nests can be found along the upper Sacramento River. Fewer individuals appear at Big Bear Lake (San Bernardino County), Lake Cachuma (Santa Barbara County), Lake Tahoe (El Dorado County), and at rivers and associated lakes and reservoirs up and down the western Sierra foothills. They are locally common in summer (from the end of March) around Mono Lake (Mono County).

SIMILAR SPECIES: The distinctive plumage pattern makes this species hard to mistake for another. At a distance in flight, it could be confused with a large gull.

STATUS: A California Species of Special Concern.

REMARKS: A program to reintroduce the Osprey to the Channel Islands began in 2000 with wild nestlings taken from Eagle Lake (Lassen County) and hacked from a release tower on Santa Catalina Island. Natural repopulation was considered unlikely in part because the nearest breeding populations (in the southern Sierra Nevada and Baja California) appear nonmigratory (Henny and Anthony 1989) and do not move far enough to reoccupy the historic nesting areas on the islands.

Kites

Of the five species of kites found in the United States, only one is commonly seen in California, and one other very rarely visits here. Distributed nearly worldwide, kites are an extremely diverse group, ranging in size from that of a robin to that of a large buteo. Most feed on prey that is relatively easily caught; some are largely scavengers. Only one species of the tropics is a swift hunter of bats and small birds in the air. No surprise then that most have small feet and beaks, although in a very few species the beak has been conspicuously modified to extract snails from their shells. Body shapes and proportions can vary greatly, and some species' build is identical to that of accipiters, perhaps to frighten off other birds pursuing the same insect prey.

WHITE-TAILED KITE
Elanus leucurus

(Pl. 2; Figs. 7, 52)

···· Wanderers

IDENTIFICATION: Crow sized, with a white head, white tail, black shoulders, and black "eyeliner" at close range. From a distance this bird looks wholly white. When perched, it appears almost legless because of its short tarsi, and it often pumps its tail up and down slowly. The wingtips reach nearly or quite to the tip of the tail. The head is big, and the beak is short.

In the adult, the eyes are bright scarlet. The back and folded wings are pale gray except for the black shoulders. The underparts are white. The white tail has gray central feathers. The sexes are alike.

The juvenile has a brown gray back and scapulars with the feathers broadly edged in white. The breast has a broad, diffuse light rufous band and large spots. The eyes are yellow brown, and the tail has a dark subterminal band.

FLIGHT: The White-tailed Kite frequently hovers and has a buoyant flight. The wingbeats are rather soft unless the hawk is headed into a wind. Gliding and soaring, the wings are usually held upward in a dihedral. The wings are pointed and usually have a black patch on the otherwise white underwing coverts at the hand. The primaries are dark, and the secondaries are light. The neck is short.

VOICE: This species is moderately vocal throughout the year; among the calls are a plaintive, repeated "glyp" when alarmed and a querulous drawn out squeal as it mobs another raptor. A rattling, chattering call is often given upon seizing prey and also when making midair talon contact with another kite.

FEEDING: The White-tailed Kite has been called "the better mousetrap," with good reason; its diet almost entirely comprises mice and voles, and it seems to catch these rodents with little difficulty. While foraging, it typically flies about fairly slowly in arcs and circles, swinging its body suddenly upright to hover, usually hovering 15 to 30 m (about 50 to 100 ft) above rodent-concealing cover.

Fig. 83. White-tailed Kites typically place their nests toward the very top of the tree. The pine shown here died after the nest was used.

The wings move more slowly than those of a hovering kestrel, and the bird appears quite clumsy. When prey is sighted, the hawk at first descends rather leisurely, and then with increasing speed. The latter part of the descent, at least, is headfirst, with the extended wings held above the back in a sharp V.

REPRODUCTION: In courtship, the male flies over the nest tree with quivering or fluttering wings nearly level or raised in a deep dihedral above the body and with legs hanging as he calls loudly. A variety of trees is used, with live oak perhaps the most common, and the nest is placed high in the crown on thin branches, unlike the nests of other hawks, which are usually in strong forks below or lower in the crown.

Most clutches consist of four handsomely marked eggs (formerly much coveted by egg collectors), which are incubated for about four and a half weeks, the young fledging five weeks after hatching. Usually, White-tailed Kites in California raise two broods, and the construction of a second nest may begin before the first set of young has fledged.

DISTRIBUTION AND HABITAT: Chiefly neotropical, the White-tailed Kite is widespread in Mexico and Central and South America. Elsewhere in the United States, small populations occur in Arizona and along parts of the Gulf Coast, and the species appears to be spreading from California into Oregon and Washington. Somewhat irruptive in distribution, non-breeding kites turn up in unexpected places, likely to prospect for vole-rich areas.

White-tailed Kites are found throughout California, at least as wanderers, although they normally stay out of coniferous forests. The bulk of the state's population is found west of the Sierra Nevada in lowlands and foothills, where they are often seen year-round.

It sometimes appears in habitats unusual for this species as it seeks rodent-rich land. It frequently takes advantage of marginal habitats, such as the edges and medians of freeways. Marshy areas, grassland, and savanna are preferred, and this bird sometimes forages in remnants of such habitats between subdivisions (see fig. 52). It commonly perches on treetop twigs, wires, and fence posts.

Coyote Hills Regional Park (Alameda County) is an excellent place to see this raptor, as is the bay shore in Albany (Contra Costa County) and Gray Lodge Wildlife Area (Butte County). It can also be found reliably in El Dorado Regional Park (Los Angeles County) and at Bolsa Chica Reservoir near Newport Beach (Orange County). Wanderers can turn up almost anywhere, including in chaparral, the mountains at 1,800 m (6,000 ft), and deserts.

SIMILAR SPECIES: A Mississippi Kite *(Ictinia mississippiensis)* is darker overall, with dark underwings and tail. An adult male Northern Harrier *(Circus cyaneus)* has a white rump and dark trailing edges on its long but rounded wings. Falcons may have a similar shape in flight but have broader bodies and a generally snappier flight. The White-tailed Kite is also somewhat gull-like in flight, but gulls are longer winged and shorter tailed and have long, projecting beaks.

STATUS: Prior to the 1940s, this species was alarmingly reduced in numbers because of shooting, destruction of foraging habitat, and perhaps downturns in the cycles of rodent populations. It has staged a comeback in recent years and is not a listed species; however, in some areas (such as San Diego County), its range is contracting because of urbanization.

REMARKS: White-tailed Kites are gregarious; several may perch on adjacent fence posts, and they may come together in communal roosts numbering in the hundreds. They may be seen at dusk flying at speed to these roosts, as far as 50 km (30 mi) away from their foraging areas, with wings angled back and the wingbeat fast but still not as snappy as that of a falcon of comparable size. The largest roosts are formed in fall and winter. Ten roosts studied in the Sacramento Valley were evenly spaced in winter; in spring, scattered smaller roosts nearer to nest sites were used. Other species roosted with this species, including crows, herons, and Northern Harriers (Erichsen et al. 1995). It will sometimes join a kettle of vultures, taking advantage of a thermal to soar.

The White-tailed Kite is preyed upon by the Prairie Falcon *(Falco mexicanus),* perhaps as the unintended result of an attempted robbery gone bad, and it is eaten by the Golden Eagle *(Aquila chrysaetos),* Red-tailed Hawk *(Buteo jamaicensis),* and Great Horned Owl *(Bubo virginianus),* especially as clumsy fledglings and during the winter months when rodents may become unavailable for the larger raptors.

MISSISSIPPI KITE   *Ictinia mississippiensis*
(Pl. 2)   `RARE`

IDENTIFICATION: A crow-sized, stocky hawk with long wings that, on perched birds, reach beyond the tip of the tail, which is square or slightly notched. The adult has a white to pale gray head, with a black spot in front of the eye and black encircling the eye. The back and folded wings are dark blue gray to brown gray, the tops of the whitish secondaries forming a narrow bar above the blackish primaries, and with a white trailing edge on the spread wing. The inner primaries on the spread wing are sometimes rufous. Underparts are gray. The tail is black. The beak is black, the eyes red, and the legs deep yellow. The sexes are similar, but the female has a somewhat darker head and inconspicuous barring on the undertail coverts.

The juvenile has a gray brown head with streaks and a short but broad whitish superciliary. The back and folded wings are dark brown, the feathers having white to buffy margins, with rows of white spots on the scapulars. The flight feathers are blackish and variably white tipped, the tips of the secondaries forming a white line. The underwing coverts are mottled reddish brown, the inner primaries showing limited barring from below. The underparts are whitish with conspicuous, wide brown streaks. The dark brown tail has narrow white bands.

FLIGHT: This species has buoyant flight with soft wingbeats. It soars often on flat wings. Although it is definitely falcon shaped with its pointed wings and long tail, the easily seen outermost primaries of this kite are far shorter than the other outer ones, unlike those of a similar-sized falcon.

VOICE: A high-pitched two-syllable whistle accompanies parts of courtship; it changes to a more chittering call when this kite is agitated.

FEEDING: Still-hunting from perches and aerial pursuit from a soar are the preferred hunting techniques of the Mississippi Kite. Foods include in large part various insects, such as locusts, beetles, and grasshoppers, and also frogs, lizards, snakes, small birds, and mice, some of these being taken in numbers by individual specialists.

REPRODUCTION: This species has never been observed to breed in California. In its areas of distribution, it is very gregarious and often nests in colonies, placing the nests in tree crotches.

DISTRIBUTION AND HABITAT: Breeding occurs chiefly in southern parts of the United States as far west as central Arizona; the range appears to be expanding. Most birds winter in South America. Vagrants may wander far from breeding areas in spring and summer.

As wanderers, Mississippi Kites have appeared in California from Humboldt to San Diego Counties along the coast and in the Coast Ranges, as well as east of the Sierra Nevada from Mono County southward to Death Valley (Inyo County), mainly from mid-May through June.

Within their normal range, they prefer savannas, grasslands, and the margins of woodlands.

SIMILAR SPECIES: A White-tailed Kite *(Elanus leucurus)* is white overall. A Peregrine Falcon *(Falco peregrinus)* has striking head markings, a more powerful wingbeat, long outermost primaries, and never appears buoyant in flight.

REMARKS: Although very rare in California, a few Mississippi Kites show up with some regularity, creating a stir in hawk-watching circles.

Harriers

Harriers have slender bodies with rather barn-owlish faces; a ruff of feathers (facial disk), which presumably augments and funnels sound, surrounds the face, and conceals large ear openings. They have acute hearing.

Although not the picture of speed and raptorial prowess, their rather unique style of foraging serves them well. Characteristically, they glide close to the ground at moderate speed while foraging. The rather blunt, long wings of harriers seem suitable for sustained foraging flight and the long tail for quick maneuvering, their slight bodies facilitating rapid pivots. Their long legs are useful for snatching prey in tangled vegetation.

North America has only one representative of this cosmopolitan group of medium-sized hawks, and it is the most widely spread of the harriers worldwide.

NORTHERN HARRIER   *Circus cyaneus*
(Pl. 2; Figs. 84, 85)

IDENTIFICATION: Like a slender, small Red-tailed Hawk with a long tail and long legs. The tips of the folded wings do not reach the tailtip. The head has a facial ruff (bordered by lighter feathers in the female and juveniles) and a conspicuous pale or whitish crescent or wedge where the cheek borders the eye. The eyes are yellow, except in the juvenile. Legs and cere are yellow. The neck looks fat, as thick or even thicker than the head. All have a white rump.

The adult male Northern Harrier has a pale to dark gray head and upper breast; the crown, back, and folded wings are darker gray mottled with brown. The tail and wing bars are indistinct when the bird is perched. In flight, the very dark tips of the primaries and dark trailing edges of the secondaries are conspicuous both from above and below.

The adult female has a brown head and a buffy, broadly streaked neck. The back is brown, the folded wings are brown with a wide bar, mottled buff. Underparts are pale buff with dark brown streaks. From below in flight, the wing and tail bars are conspicuous, as are the broad dark trailing edges and dark-

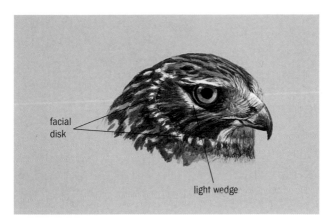

facial disk

light wedge

Fig. 84. All harriers have a facial disk, often less obvious on the male. Note the facial disk and light wedge.

tipped primaries of the wings; the secondaries are, overall, darker than the primaries. The dark head and neck contrast with the pale breast and belly.

The juvenile is much like the female, except the underparts are dark rufous buff, fading with age, and thinly streaked on the breast or not at all. The eyes are brown or gray.

FLIGHT: The Northern Harrier is most often seen cruising low over fields and wetlands, in typically unhurried buoyant flight, sometimes rocking from side to side, especially in false starts after prey. The white rump is very conspicuous when the upper side is seen. The tail is long. The bird appears light bodied and has soft, rowing wingbeats. The wings are held in a strong dihedral in active flight, in a slight dihedral or flat while soaring; they are narrow but not very pointed, preventing Northern Harriers flying overhead from being confused with falcons.

VOICE: This harrier produces a variety of calls, all almost exclusively during the breeding season. Staccato chuttering is a common vocalization directed at humans or predators near the nest.

FEEDING: The foraging flights of the Northern Harrier often appear erratic and indecisive until you realize that the hawk is actually seeking out promising tufts of cover likely to hold prey. After a few wingbeats, the Northern Harrier sails along on wings that

Fig. 85. This moderate-sized brown hawk with a prominently barred tail and fairly long, somewhat narrow wings is recognized instantly as a Northern Harrier (a juvenile in this case) because of its conspicuous white rump.

angle back and also form a dihedral. A harrier appears tireless in its search. When prey is spotted, the hawk drops or turns sharply downward on flaring wings and tail. Surprisingly, it is also capable of some speed and can catch small birds in a tail chase. A born opportunist, the Northern Harrier feeds on a wide variety of animals—grasshoppers, frogs, ducklings, and small pheasants, although it is essentially a vole specialist. It also eats carrion at times.

REPRODUCTION: The courtship flight of the Northern Harrier is spectacular. The male, and sometimes the female, rises to a height of several hundred feet. It then dives, half-folded wings still beating, and pulls up again, only to dive once more, in an undulating pattern. Frequently, the bird turns upside down as it starts into the next dive. The display is often accompanied by calls.

The nest is on the ground, an unusual but not unique location for a hawk nest. The female lays typically four to six eggs, although as many as 10 have been observed. For such a big hawk, these are large clutch sizes, perhaps reflecting the need of the species to compensate for the loss of entire nests to ground predators. Both the male and female, when intruders get too close to their nest, fly overhead, calling in alarm. When the male brings

Fig. 86. Northern Harrier nest site, in pickleweed (*Salicornia* sp.).

food to the nest, the female usually meets him and flies beneath
him, whereupon he drops the prey to her in a midair transfer, as
he also does during courtship. Northern Harriers tend to nest in
loose colonies, especially if voles are plentiful, and the males may
indulge in bigamy and even polygamy, raising broods with more
than one female (sometimes five or more).

The eggs are incubated for four and a half weeks, and the
young fledge after about five weeks.

DISTRIBUTION AND HABITAT: The Northern Harrier is widely distributed across the northern Holarctic, with some populations
wintering in tropical and subtropical regions.

Widespread in California, this species has become uncommon in the southern part of the state. California populations are
greatly augmented by winter migrants; during migration this
hawk can be seen almost anywhere, even above timberline.

The Northern Harrier seems to prefer wetlands, both freshwater and salt marshes, but it is also commonly found over grasslands, cultivated areas such as alfalfa and wheat fields, including
cultivated areas in the Mojave Desert, and even sagebrush such as
in the Great Basin. The species is readily seen on coastal grasslands and in the Central Valley.

It often perches on the ground or on fence posts.

SIMILAR SPECIES: When perched on the ground or a fence post,
the female and juvenile Northern Harrier might be confused
with a Red-tailed Hawk *(Buteo jamaicensis)* (look for the har-

rier's pale wedge below the eye) and the male with a White-tailed Kite *(Elanus leucurus)* (it lacks the latter's black shoulders).

STATUS: A California Species of Special Concern and a federal Bird of Conservation Concern outside California.

REMARKS: These harriers are usually shy, not allowing close approach. Wintering harriers in California, as elsewhere, are normally tolerant of one another, especially if food is abundant, although they may fight over a large food item such as a dead coot. They sometimes come together in communal night roosts where as many as several hundred may meet to spend nights together on the ground.

Accipiters

California is home to three species of accipiters; although differing in size, all three have essentially the same proportions and are not always readily told apart.

All three pursue their prey with great verve and agility, often in cover, and their hunting activities may frequently go unnoticed. Although it is generally agreed that accipiters are sprinters and seem to lack the staying power of falcons, all, especially the goshawk, at times chase their quarry over great distances and even up into the air, much as a falcon does. Accipiters are also agile and speedy runners, particularly so the smaller species, and commonly pursue their prey on foot into thick vegetation.

At close quarters, the brilliant red or yellow eyes of most of these hawks are a conspicuous feature, and at times the bright yellow cere can also be very prominent.

SHARP-SHINNED HAWK   *Accipiter striatus*
(Pl. 3; Figs. 6, 87)

IDENTIFICATION: A very small hawk, smaller than a pigeon but bigger winged and longer tailed, with spindly "toothpick" legs. Although the sexes are alike in color and pattern, the male is often tiny, substantially smaller than the female. The size difference is more pronounced in Sharpshins than in most other hawks. The tail, both perched and in flight, is often (but not always) square at the tip, with the outer tail feathers nearly equal in length to the other rectrices. The short wings do not reach midtail when the bird is perched. This hawk has a small round head.

The adult's head has a dark (but not black) cap, rusty cheeks, a white throat, and red eyes narrowly ringed with dark feathers. The back and folded wings are blue gray. White spots on the back can be nearly or completely covered. Underparts are white, prominently barred rufous, the barring occasionally so extensive that it obscures the white, especially on the leg feathers and in males; the undertail coverts are white. The gray brown tail has three or four conspicuous, wide, black brown bands, which appear narrower when seen from below, except for the terminal band. The cere and legs are yellow.

The juvenile's head is brown with variable buff or white streaking, pale superciliary, and rufous brown cheeks. The eyes are pale to deep yellow and staring, giving the young Sharpshin a rather startled look (see fig. 87). The back and folded wings are brown, the feathers narrowly edged rufous, the scapulars with scattered white spots that are not always visible. The throat is white with fine median streaks, and underparts are white with broad reddish brown or, sometimes, narrow, dark brown streaks that become bars on the flanks. Undertail coverts are white.

FLIGHT: The head is small and round and the neck short, while the tail is long but often appears relatively shorter than in other accipiters. In point-to-point flight, the Sharpshin alternates several hasty wingbeats with glides on set wings. In a soar with fully spread wings and tail, the barring of the flight feathers is conspicuous. The short, rounded, rather broad wings are often held "hunched" forward instead of straight across from tip to tip, and the small head seems recessed. When soaring in a wind, a Sharpie seems to get blown about like a dead leaf, but can stoop effectively. It makes very small circles when riding a thermal.

VOICE: The voice of this hawk is appropriately reminiscent of that of a small bird—high pitched and not very powerful. The most common vocalization is a staccato series of "ki-ki-ki," which is given in the presence of potential nest robbers and also sometimes during migration while harassing another, larger raptor such as a Red-tailed Hawk *(Buteo jamaicensis).*

FEEDING: The Sharpshin is an expert ambush-hunter and quickly discovers the drawing power of backyard bird feeders on songbirds. The little hawk dashes from cover at blinding speed and snatches a finch or sparrow as it rises, and this hawk can turn on the proverbial dime. It also undertakes speculative flights, skimming low over hedges, fences, and houses in hopes of surprising quarry unawares. The Sharpshin is generally more aerial than its close relative, the similar Cooper's Hawk *(Accipiter cooperii),* and commonly pursues birds into the air or stoops at them from above.

The Sharpie is also fleet of foot and at times chases its quarry by running after it into bushes. This little hawk's boldness is legendary; it has been known to enter houses through open windows in pursuit of exotic morsels such as parakeets.

Nearly all prey is avian, for the most part small- to midsized birds such as robins, a common prey of the female. The tiny male tends to feed on sparrows, chickadees, and other backyard favorites; even hummingbirds may appear on the menu (Peeters 1963b). Rarely, birds as large as pigeons and quail are taken, along with very small numbers of mice, lizards, frogs, and insects.

The Sharp-shinned Hawk maintains a very high metabolic rate, and thanks to its capacious crop, it can pack away prodigious amounts of meat from a large kill. A truly sated Sharpie looks extremely top-heavy, as if it has had a silicone implant (see fig. 6).

REPRODUCTION: Courtship flights include high soaring and calling, undulating roller-coaster flights by the male (his white undertail coverts projecting out the sides), and tandem flying in a straight line by the pair. It is not uncommon for a Sharpshin to breed when less than a year old, while still in the juvenal plumage.

The stick nests, often surprisingly small, are built in dense groves of usually midsized conifers, in the tops of live oaks, and sometimes in deciduous trees. They are generally well hidden, the only clues being the hawks' calls, possibly molted feathers, and the butcher block of the male, where he plucks prey before delivering it to the female. Nest groves are usually on hillsides or hilltops, close to grass, often near chaparral-covered slopes, but typically not near water as are those of the Cooper's Hawk.

From two to six eggs are laid, most commonly four, a relatively high number that reflects the high mortality of this small active raptor. Incubation lasts about four and a half weeks, and the young leave the nest from three to four weeks after hatching.

DISTRIBUTION AND HABITAT: The Sharp-shinned Hawk is chiefly a boreal breeder in North America, with populations extending down both coasts and into the interior, as well as in the Rocky Mountains and coniferous areas of Mexico and Central America (although southernmost birds may be a different species).

Fig. 87. Juvenile Cooper's Hawks are characterized by a full nape, an often elongated head, and a fierce look. Juvenile Sharp-shinned Hawks have a rounder head, a very short beak, and a startled expression.

For breeding habitats in California, the Sharpshin seems partial to mountain coniferous forests but also breeds, perhaps in equal numbers, in coastal coniferous forests and in live-oak woodlands of the Coast Ranges.

Every fall, migrants coming from other states (including New Mexico [Hoffman et al. 2002]) and from Canada greatly increase the numbers of Sharpies in California; the migrating hawks pass through or winter here. Sharp-shinned Hawks are found throughout California in winter.

This species, which by build and inclination is a hawk of forest and woodland interiors, sometimes follows its prey species, sparrows of various kinds, into open grasslands and deserts, and into cities and just about every habitat in between. It is found more often in open country than is the Cooper's Hawk.

The Sharpshin is perhaps most often seen flying high over hilly terrain on migration (it is one of the most numerous species seen at migration observation points) or raiding a backyard bird feeder.

SIMILAR SPECIES: Females are sometimes confused with male Cooper's Hawks, whose size they approach. In flight, the Cooper's has longer and narrower appearing wings, a longer neck, and particularly as a juvenile, a longer tail (see the "Similar Species" section in the Cooper's Hawk species account). When seen close up, the eye of the adult Sharpie is narrowly ringed with dark feathers (absent in the adult Cooper's Hawk), and the eye of the juvenile Cooper's Hawk appears more heavily browed, which gives it a fierce expression (the dark eyes of the adult often obscure this difference). A tiny, male Sharp-shinned Hawk might be mistaken for a nonraptor from afar when perched, a misperception that is soon dispelled by the much larger and barred wings when it takes flight.

STATUS: A California Species of Special Concern.

REMARKS: The male Sharp-shinned Hawk is the smallest of the raptors found in California. The Sharpshin has the longest legs of all the accipiters, relative to size, and can look very fragile, especially when seen close up. Not surprisingly, it is sometimes killed by other raptors, including Peregrines *(Falco peregrinus)* and even Bald Eagles *(Haliaeetus leucocephalus).*

Appearing high-strung and very active, they are dashing little raptors that are sometimes loathed by people who feed songbirds and who do not understand that the hawks are most likely to pick off individuals weakened and slowed by disease, physical flaws, or a lack of alertness. First-year Sharpies that have not yet honed their skills annually die by the hundreds in the winter months because of their failure to make a kill in the space of two or three days while already in poor physical shape. Plumage soaked from rain can make foraging impossible and has doomed many a Sharpshin. Older birds not only know how to deal with difficult conditions but also are generally heavier in weight and in much better physical condition.

Of all the hawks, the Sharpshin is the most likely to come to harm in collisions with window panes, probably because it sees its own reflection and, mistaking it for quarry, instantly veers toward it and slams into the glass.

COOPER'S HAWK *Accipiter cooperii*
(Pl. 3; Figs. 16, 18, 23, 87)

IDENTIFICATION: A medium-sized hawk, about the size of a crow, slender and often lanky-looking, with long legs and a very long tail rounded at the tip. The head frequently appears long and flattish because of the long nape feathers and the raised feathers at the back of the head (the occipital crest); the eye appears to be placed far forward. The wingtips fall far short of half the tail length when the bird is perched.

May breed here.

The adult's head is black capped, with rufous brown cheeks and orange to red eyes. Upperparts and folded wings are blue gray (or brown gray in young adults). Underparts and underwing coverts are white with heavy, bright, rufous barring that may in places fuse into solid color, especially on the sides of the breast. Undertail coverts are white. The tail is brown gray with three or four broad, widely spaced, blackish bands, narrower from below. Seen from the back in good light, the reddish nape often clearly separates the black cap from the back, especially when the crest is raised.

The juvenile has an inconspicuous light or no superciliary; the head is brown capped, with dark brown streaks on a buffy nape. The folded wings and upperparts are dark brown, the scapulars with scattered white spots. The upper breast is often buffy and always more heavily streaked than the white belly (which may be almost clear), so the underparts appear two-toned at close range. Undertail coverts are pure white, sometimes with fine streaks. The whitish leg feathers have conspicuous brown diamonds. When the juvenile is in flight, the remiges are boldly barred, with clearly defined dark subterminal bands along the trailing edges of the wings. The tail is banded as in the adult, but it is longer and often spatulate, giving a "panhandle" impression.

A very rare as yet unnamed form has a gray breast in the adult, and gray breast streaks and back in the juvenile (Jesus et al. 1995).

FLIGHT: A series of floppy but quick wingbeats alternates with glides in active flight. In point-to-point flight, this species flies very straight, neither rising nor dropping. The rather long and narrow appearing wings are nearly straight across the leading edges when spread in a soar, with the head and neck projecting far forward. The tail is often conspicuously white tipped.

VOICE: The Cooper's Hawk, like other accipiters, has always been regarded as a rather quiet hawk, but recent studies and its nesting in cities in recent years have revealed that it is in fact very vocal during the breeding season. Most vocalizations have a distinctive reedy quality. Male and female sing duets at daybreak and produce a great variety of calls for occasions such as greeting each other as the season progresses. Alarm calls around the nest are a series of staccato "ca-ca-ca" calls, and the begging notes of the fledged young are soft whistles.

FEEDING: A Cooper's Hawk hunts by ambush and also by specula-

Fig. 88. An urban Cooper's Hawk nest in a Berkeley neighborhood.

tive flights intended to surprise vulnerable prey. This raptor is exceptionally fleet of foot and may run after potential quarry into thick poison oak tangles and other shrubbery. It may even slap bushes with its wings, trying to evict the quarry. The hawk is acutely aware of the meanings of sounds, such as the call of a quail. Occasionally, it pursues birds high into the air or attacks in a stoop, although less commonly than does a Sharp-shinned Hawk *(Accipiter striatus)*.

An urban Cooper's Hawk feeds mainly on Mourning Doves *(Zenaida macroura)*, American Robins *(Turdus migratorius)*, House Sparrows *(Passer domesticus)*, and city pigeons. Away from cities, midsized birds such as Scrub Jays *(Aphelocoma coerulescens)* are commonly taken, along with quail, woodpeckers, young cottontails, and lizards. One nest contained the remains of three adult Western Screech Owls *(Otus kennicottii)*.

One juvenile, frustrated at not being able to reach a quail in a bush, began to utter begging calls. The Cooper's Hawk can be very dashing however; an adult male flew at telephone pole height for about 180 m (600 ft), accelerating as it went, then climbed steeply to about 50 m (150 ft) to intercept and catch a Brewer's Blackbird *(Euphagus cyanocephalus)*.

The Cooper's Hawk likes to eat in private and usually carries or drags prey under a bush or overhanging tree branches.

REPRODUCTION: The male flies with stiff, deep, slow wingbeats in courtship display while keeping the tail folded, the white undertail coverts sticking out from the tail base like a tutu. Sometimes both sexes participate in such flights. Undulating flights with shallow dives are also seen. Prior to nest building, a female Cooper's Hawk may spend hours quietly sitting in plain view on high perches, unusual behavior for an ordinarily secretive species.

For nest sites, Cooper's Hawks in central and southern California prefer canyon bottoms or hillside benches in fairly dense stands of oaks, nearly always by a stream or pond or even a temporary pool. They like nearby chaparral for foraging. In southern California, they also nest in olive orchards and eucalyptus trees. Conifers are used in the Sierra Nevada and sometimes elsewhere. In cities, nests are placed in park, backyard, and even street-side trees.

Often the nest is well concealed in a dense crown, and usually very near old nests from previous years. The male defends the nest area against other Cooper's Hawks, most strongly so against other males (Boal 2001).

From three to six eggs are laid, four being common; they are incubated for about four and a half weeks. The young leave the nest at four to five weeks.

DISTRIBUTION AND HABITAT: The Cooper's Hawk breeds from southern Canada and most of the United States into northern Mexico, although it is largely absent from the northern Great Plains.

While on migration or wintering, this species is found throughout California except for the snowy heights of the Sierra. For nesting, however, it is more particular, preferring riparian and oak woodland and, in recent years, ornamental tree plantings (such as elm and eucalyptus) in some central California cities. This species can be found on occasion nesting in montane forest, but the main breeding populations are probably in the Coast Ranges and in the foothills surrounding the Central Valley, where they are locally numerous.

SIMILAR SPECIES: A Sharp-shinned Hawk in flight has shorter, more rounded and broader appearing wings and often a square tail that also seems short for an accipiter; if the tail is rounded, it could be a Sharpie or a Cooper's. If the head of a soaring bird is recessed relative to the leading edge of the wings, it is certainly a Sharpie. If the wings are straight across with the head projecting, it could be either a Sharpie or a Cooper's, and tail length and other criteria must be used for identification. A Northern Goshawk *(Accipiter gentilis)* has pointier wings and is much larger. Close up, a Cooper's Hawk's eyes are not dark ringed like a Sharpie's. For close-up comparison of Sharp-shinned and Cooper's Hawks, see fig. 87.

STATUS: A California Species of Special Concern.

REMARKS: Although an adult Cooper's Hawk appears compact and powerful, especially a female, juveniles look lanky and floppy. Their tails seem extravagantly long, the wings often hang loosely from the sides, and the plumage is lax and soft. The extended head and full nape are often conspicuous.

NORTHERN GOSHAWK   *Accipiter gentilis*
(Pl. 3; Figs. 10, 24,36)

IDENTIFICATION: A large hawk, the size of a Red-tailed Hawk, chiefly of mountain coniferous forests and aspen woodland. The male is somewhat smaller than the female. The tips of the folded wings reach halfway to the tip of the tail. The tail is rounded at the tip. The legs are short, and the legs and cere are yellow.

very
rarely seen

The adult's head has a black crown and a broad black stripe from eye to nape; wide white superciliaries extend to the nape and may isolate the crown. The eyes are deep yellow, deep ruby red, or very dark red brown. The back and folded wings are blue gray. Underparts appear silvery because of fine, black brown bars and vermiculations on white, with scattered, dark shaft streaks. The blue gray tail has three or four incomplete black bands. Undertail coverts are white. Underwing coverts are marked like the breast; the primaries are strongly banded, but the secondaries are only faintly so. From above, the spread wings show brown primaries, with all other feathers blue gray.

The juvenile's head is beige with variably fine streaks and a conspicuous, broad, pale superciliary. The eyes are variably yellow. The back and folded wings are brown and beige, strongly mottled in appearance; the neck is pale rufous with dark, conspicuous streaks. Underparts are light colored, broadly streaked with dark brown; the leg feathers have dark diamonds. The tail is tan with three or four wide, wavy, dark brown bands that are narrowly bordered with beige. The whitish undertail coverts are streaked, often broadly. The flight feathers are conspicuously barred on the underside.

FLIGHT: A Northern Goshawk alternates several rapid wingbeats with gliding. It appears powerful in flight and often looks chesty. It soars with wings flat. The tail is rounded or wedge shaped, with conspicuous, wide, dark bars.

VOICE: This hawk is generally silent except during breeding season, when it employs a great variety of vocalizations, including a wail somewhat like that of a gull, calls given in the form of a morning duet, and alarm and nest-defense calls, which are a staccato series of high notes, rather like a heavy-duty version of the Sharp-shinned Hawk's *(Accipiter striatus)* alarm calls.

FEEDING: The Northern Goshawk forages by still-hunting, frequently moving to new perches, and by speculative search flights, suddenly popping over prey-concealing cover. It is a persistent chaser, zigzagging after quarry between sage bushes, but it also crashes into dense cover after prey and sometimes pursues it high into the air and over distances exceeding a kilometer. Occasionally, one can be seen rocketing down a mountainside in pursuit of a grouse.

Prey animals are almost exclusively midsized mammals, such as rabbits and ground and tree squirrels, and midsized to large birds, from American Robin *(Turdus migratorius)* to Blue Grouse *(Dendragapus obscurus)* in size. One juvenile caught and ate a full-grown hen Wild Turkey *(Meleagris gallopavo)* (Golet et al. 2003). Mice and shrews are sometimes also taken, along with small birds, when the opportunity presents itself, and nestling birds are a frequent prey brought to nestling goshawks. Reptiles and insects are rare prey items.

REPRODUCTION: The Northern Goshawk displays in various ways

Fig. 89. An adult Northern Goshawk on its nest (arrow).

during courtship; both male and female may fly together with slow, deep wingbeats and, during intermittent glides, hold their wings in deep dihedrals. The brilliantly white undertail coverts are flared out sideways. High circling while displaying these coverts also serves as territorial announcement.

In the Sierra Nevada, it commonly builds its nest in dense, large stands of lodgepole pine *(Pinus contorta)*, usually a few yards from a meadow or road, which is used to facilitate approach. It hunts the surrounding forests and meadows. The proximity of a creek appears less important than it is for the Cooper's Hawk *(Accipiter cooperii)*, although water is frequently nearby. It may also use other species of pines in smaller groves and, on the eastern side of the Sierra, quaking aspens *(Populus tremuloides)* along streams, sometimes far from conifers; here the hawk forages on sage slopes. In the northwest, it nests in groves of old-growth Douglas-fir *(Pseudotsuga menziesii)*. In north-central California, this species now nests in older second-growth forests, especially those that contain remnant groves of first growth. Very rarely, the nest tree is a solitary ponderosa or Jeffrey pine *(Pinus ponderosa, P. jeffreyi)*.

The nest is nearly always placed against the trunk under or in the lower part of the canopy, so it is quite visible from below. It is usually built in the vicinity of old nests.

From two to four eggs are laid, commonly three, and they are incubated for four and a half to five weeks. The young leave the nest after five to six weeks.

DISTRIBUTION AND HABITAT: The Northern Goshawk is a circumpolar breeder of the Holarctic. In North America it breeds as far south as the western mountains of Mexico.

This species can be found throughout California, but in most places it is very rare and seasonal, irregular in yearly occurrence, or all three. It is nowhere common in the state.

Breeding activity is confined to the species' summer habitats, chiefly subalpine forest above 1,800 m (6,000 ft) and at less than half that elevation in the coastal mixed coniferous forests of the northern Coast Ranges; a few pairs breed in the higher mountains of southern California and in aspens in the eastern Sierra Nevada.

Resident Northern Goshawks may move down to lower elevations during the winter months, and migrants augment the state's population. Although not as conspicuously as in more northerly states, substantial numbers of goshawks occasionally enter California during so-called invasion years. On migration, a few individuals may turn up in oak woodlands, marshes, agricultural lands, and even cities and deserts.

SIMILAR SPECIES: Both the Cooper's and Sharp-shinned Hawks are much smaller than a Northern Goshawk. A juvenile Red-shouldered Hawk *(Buteo lineatus)* is shaped like an accipiter, approaches a Northern Goshawk in size, and has a similar flight style. In flight, a juvenile may resemble a juvenile Red-tailed Hawk *(Buteo jamaicensis)*, and an adult, with its long and slender wings, may resemble a Gyrfalcon *(Falco rusticolus)*.

STATUS: A California Species of Special Concern.

REMARKS: Although the Northern Goshawk breeds principally at

Fig. 90. A stack of Harris's Hawks, an unusual association among raptors.

HARRIS'S HAWK

Parabuteo unicinctus

(Pl. 4; Fig. 90)

RARE

IDENTIFICATION: A large, Redtail-sized hawk. The wingtips of a perched bird reach halfway to the tip of the tail. The tail is long and full. The long legs, cere, and face are yellow. Sexes are alike in color and pattern, the female larger.

The adult's head and body are dark brown, the shoulders and leg feathers rufous. The tail is black with a broad white base and a conspicuous white tip. In flight, the rufous underwing coverts contrast with the dark gray undersides of the flight feathers. Tail coverts are white.

The juvenile much resembles the adult, except that the undersides are variably streaked white and black brown, sometimes so heavily that the underparts appear nearly solid dark. On lighter individuals, heavy streaks on the upper breast form a bib. Leg feathers are rufous or rufous barred. In flight, the tail and remiges have numerous narrow pale bars without wide terminal bands on the tail or dark trailing edges on the wings.

FLIGHT: This hawk flies with fast wingbeats, and soars on flat wings. The wings are long but rounded at the ends.

VOICE: Most vocalizations are generally rasping calls of some sort, from soft shirring greeting notes to a harsh, coarse "grrahh" uttered in nest defense. These are very unlike the calls of any other hawk and therefore not readily recognized as coming from a raptor; rather, some resemble the shout of rage of a human having hit a thumb with a hammer. It also has, however, a rapidly repeated "keking" call that seems to be given in the presence of a mammalian predator.

FEEDING: The Harris's Hawk still-hunts but does not appear to stay on a given perch very long, moving frequently to new lookout sites instead. These raptors may hunt singly, in pairs, or even in groups of three or six or more birds, not all of them necessarily related to one another. They like to touch, and on occasion they even perch on one another (sometimes three deep for more than 5 minutes [Dawson and Mannan 1991b]) while looking for prey, a remarkable display of mutual tolerance (see fig. 90).

For a hawk that can appear so sluggish as it rests on exposed perches, the Harris's Hawk turns into a vigorous, agile predator when in pursuit of prey, and a group of these raptors ganging up on a large item like a jackrabbit resembles a rugby team at scrimmage. Cooperative hunting involves tactics such as flushing and relay chases and can result in taking quarry usually too large for a single bird, which is then shared fairly peaceably; the optimal group size is five (Bednarz 1987). This predilection for social hunting has been discovered by falconers, who find that Harris's Hawks not only seem to understand the function of man and dog as game flushers but also appear to enjoy the company of their handlers and, sometimes, the canines. They can be flown as teams in captivity as well.

Wild individuals take a variety of prey, chiefly mammals and birds, besides some reptiles and invertebrates.

REPRODUCTION: Courtship, too, can be a social affair, with as many as eight birds soaring, calling, and diving. Males have been seen to stoop vertically from up to 200 m (600 ft), pulling out at the bottom of the dive to land on the perched female's back to copulate with her.

Harris's Hawks not uncommonly maintain a ménage à trois with two males, one dominant over the other, joining with the same female and aiding her in nest building, incubation, and feeding of the young (Dawson and Mannan 1991a). They use other unconventional mating systems as well, all of them in North American populations. South American Harris's Hawks behave like proper Methodists.

Ironwood *(Olneya tesota)* and palo verde *(Cercidium* spp.) are among the trees used for nesting; this hawk sometimes also nests on human-made structures.

Incubation lasts about five weeks; the young fledge in about six weeks. Sometimes two or even three broods are raised in one nesting season, and the offspring from the earlier nestings may help in the feeding of their younger siblings.

DISTRIBUTION AND HABITAT: In California, the range of the Harris's Hawk is restricted to the south. Outside California, it occurs from southern Arizona and New Mexico and central Texas southward through Central America and nearly the length of South America.

This hawk was extirpated in California as a breeding bird in the 1950s, and a reintroduction in the 1980s apparently succeeded only briefly. However, there are periodic incursions of this species from its principal range to the east and south, with the hawks appearing as far north as Victorville (San Bernardino County) and breeding once again in the state. These periodic occurrences apparently result from the production of large numbers of young under optimal conditions (i.e., a large prey base resulting from unusually high rainfalls) in their normal ranges, which push "surplus" birds into our state (Patten and Erickson 2000).

The typical habitat of this hawk includes dry savanna, desert riparian woodland, and scrub.

SIMILAR SPECIES: Juvenile Redshoulders *(Buteo lineatus)* are smaller and lack the dark, bold streaks on underparts (compared to a juvenile Harris's). None of the dark morph buteos have a tail as long relative to their wing length, nor the distinctive tail pattern.

STATUS: A California Species of Special Concern.

REMARKS: This species, because of its elaborate social system, displays a dominance hierarchy. Somewhat surprisingly, when perched upon each other, it is the subordinate individual that usually sits on the dominant one, whereas the dominant bird perches highest when separate perches are used. Such dominance hierarchies, established by fighting, reduce aggression, important in a social and heavily armed species. Larger birds and older individuals dominate smaller and younger ones, the latter recognized by their plumage (Dawson and Mannan 1991b).

Because of its popularity with falconers, a Harris's Hawk seen in the wild may in fact be an escaped or released bird. Such individuals are likely quite capable of fending for themselves and, in the south of the state, potentially augment the wild population.

This species is sometimes called the Bay-winged Hawk, a name some scientists prefer, especially for the South American race.

RED-SHOULDERED HAWK *Buteo lineatus*
(Pl. 5; Figs. 46, 91, 93)

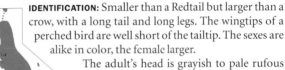

occurence
likely

IDENTIFICATION: Smaller than a Redtail but larger than a crow, with a long tail and long legs. The wingtips of a perched bird are well short of the tailtip. The sexes are alike in color, the female larger.

The adult's head is grayish to pale rufous with fine, tan stripes and an indistinct malar patch. The upper back is rufous with side dark streaks; the scapulars are blackish brown, barred whitish, giving the entire back a mottled appearance. The folded wings are deep rufous at the shoulder. Greater coverts, remiges, and tail feathers are boldly barred or banded black and white; the folded wing has a checkerboard pattern. The upper breast is bright rufous, as are the bars of decreasing width on the remainder of the underparts. The feet and legs are yellow.

The juvenile's head and back resemble those of the adult but are browner and mottled with white, the shoulder feathers brown with rufous edges. The barring on the remiges is much less crisp, brown and tan. The throat is rufous, the bib streaked blackish and bordered with rufous, and the rest of the underparts are broadly barred rufous brown. The tail is dark brown with tan bars.

FLIGHT: In active flight, short glides alternate with several quick wingbeats. A Redshoulder glides with wings bowed downward and soars on flat wings held somewhat forward. Even when the hawk is high overhead, the bright black-and-white barring of the wings and tail are striking, along with a white or pale curved "window," or crescent across the midregion, at the base of the

Fig. 91. Male Red-shouldered Hawk displaying high above his soaring mate. Note wing "windows."

outer primaries—a diagnostic feature from both above and below. In flight, the juvenile seen from above looks quite brown, with the pale crescents across the primaries prominent.

VOICE: Perhaps California's noisiest raptor, a Red-shouldered Hawk, whether perched or flying, often draws attention to itself with its loud and repeated clarion calls, especially during the nesting season. Steller's Jays *(Cyanocitta stelleri)* are frequent mimics, but their attempts pale in volume and purity of tone. The most common call is composed of two notes, the last trailing off, and is apparently always repeated at least once but usually several times. The function of this vocalizing appears to be chiefly territorial announcement, although it is also associated with courtship. Other single notes are often given in series, as for example during nest defense.

FEEDING: For a raptor that appears so lively on the wing, the Red-shoulder is amazingly sluggish when foraging; its preferred method is still-hunting. It may remain on a chosen perch for astonishingly long periods, patiently studying the ground below for signs of prey.

This hawk is a consummate opportunist when it comes to acceptable food. Although small mammals such as voles and shrews usually constitute the bulk of the diet, it often catches snakes, lizards, frogs, toads, and baby birds (along with occasional adults, although songbirds on a backyard feeder usually ignore this raptor even when it is perched nearby), and it is not above walking about on the ground after a rain, eating earthworms. This generalized diet enables Redshoulders to survive, for example, in urban habitats.

REPRODUCTION: The courtship display flights can be exuberant. The pair rises together, with the male gliding in arcs high above the soaring female, his tail folded, his wings in a strong dihedral and angled forward so that he resembles a stylized Christmas tree angel (see fig. 91). He then descends in a series of shallow dives, during which he turns on his side with quick wingbeats and eventually dangles his legs.

The nest is traditionally built in tall trees of riparian woodland, very often in western sycamores *(Platanus racemosa)* and Fremont cottonwoods *(Populus fremontii),* but in recent years, Redshoulders have developed a great fondness for eucalyptus,

Fig. 92. Nest of a Red-shouldered Hawk in a sycamore in riparian woodland. Nests are often hard to see, as are their occupants. Sometimes only the tail of the adult sticks out.

particularly the dense-crowned varieties, in urban settings. Even in riparian habitats where traditional trees are available, this hawk preferentially nests in eucalyptus (Rottenborn 2000). Street-side fan palms *(Washingtonia filifera)* are used as well.

This species lays from two to five eggs, a clutch of three being the norm in California. Incubation lasts nearly five weeks, the young fledging about six and a half weeks later. Independence from the parents follows about three weeks after.

DISTRIBUTION AND HABITAT: The California form of the Red-shouldered Hawk, which extends into Oregon and Baja California, is isolated from other forms confined to the eastern half of the United States, southeastern Canada, and central and northeastern Mexico.

Originally found in California almost exclusively in riparian woodland west of the Sierra Nevada, this hawk is presently expanding its range, but few Red-shouldered Hawks breed in any other western state. It is most common in the outer Coast Ranges but in recent decades has greatly increased in numbers inland as well (nesting, for example, in Inyo County). In the north, it has spread to the Oregon border and into Shasta County. With the exception of the open deserts and continuous coniferous forests, Redshoulders can be found mostly year-round throughout California in suitable habitat, although they are scarcer east and south of the Sierra and in the northeast, especially in winter.

Traditional riparian habitat has been severely reduced in California. Urban nest site selection suggests that the presence of a stream does not appear to be an absolute requirement; however, riparian woodlands, for example in the San Francisco Bay Area, are still good places to look for Red-shouldered Hawks. In years of vole abundance, these hawks, principally the juveniles, may be seen in open grasslands and savannas, far from their traditional habitat.

Redshoulders can be found in truly amazing numbers along Highway 1 and side roads in southwestern San Mateo County in winter, and in much of Sonoma County. As elsewhere, they can often be seen perched on telephone poles and the fat, lower cables strung between them.

SIMILAR SPECIES: Juvenile Redshoulders are sometimes mistaken for adult Broad-winged Hawks *(Buteo platypterus),* but the

Fig. 93. A juvenile Red-shouldered Hawk can greatly resemble a midsized accipiter, although the large, round buteo head usually gives it away. Cables strung between telephone poles are a favorite perch of this species.

Broadwings have uniformly dark brown backs, short legs, and fewer tail bars. In flight, a Redshoulder, with its long tail and relatively short wings (compared to a Redtail's), is very reminiscent of an accipiter, and the juvenile may be confused with a juvenile Northern Goshawk *(Accipiter gentilis)* or a Cooper's Hawk *(A. cooperii)*, even when perched. In flight, however, those accipiters' wingbeats between glides are choppier than the Redshoulder's, whose beats are more flowing.

REMARKS: Red-shouldered Hawks like more cover than do Redtails *(Buteo jamaicensis)*, especially for nesting. Where both species occur together, the Redshoulder's nest is usually more concealed, and encounters between the two, not uncommon because the territories may overlap, result in the Red-tailed Hawk chasing the smaller Redshoulder into the trees.

Perhaps the most beautiful of our hawks, the California form is more brightly colored than forms to the east. It was formerly often called the Red-bellied Hawk.

BROAD-WINGED HAWK   *Buteo platypterus*
(Pl. 5)   `RARE`

IDENTIFICATION: A big-headed, very small, plump buteo, the size of a crow. Wingtips of a perched bird reach three-quarters to the tailtip. The legs are short. A dark morph exists, entirely blackish brown with tail bars as below.

The adult's head is brown with a darker malar patch, but the head and neck have some rufous. The back and wings are brown. Variable reddish brown barring on the upper breast often forms a bib, and the remainder of the underparts is white with bold, reddish brown chevrons (extremely variable in number and extent). The tail has two conspicuous dark wide bands separated by a wide pale band.

The juvenile's head is brown, the back and wings darker brown. Underparts are white or creamy with variable streaking, from almost clear to very heavy. The tail has numerous narrow, alternating dark and light bands, the subterminal dark band the widest.

FLIGHT: The wings are moderately long and wide but pointed and are held level when soaring. In flight overhead, the dark breast and wide, dark trailing wing edges (less obvious in juveniles) contrast with the generally pale underparts. This species, in a soar, is proportioned more like a prairie-savanna buteo than a typical woodland hawk.

VOICE: Broad-winged Hawks are not likely to be heard during their journey through California. In its nesting lands, the most common call is a very high-pitched whistle reminiscent of that of a songbird such as a peewee (*Contopus* spp.).

FEEDING: A confirmed still-hunter that rarely forages using search flights or by soaring, the Broadwing watches for prey from perches under the canopy of its woodland habitat and on telephone poles and wires, usually along edges of roads and clearings. The bulk of the diet is composed of small mammals, followed by nestling and fledgling birds (during the nesting period), amphibians, reptiles, and invertebrates.

REPRODUCTION: Courtship displays are similar to those of other buteos, including soaring together and calling; these are unlikely to be observed in California. The nests are built chiefly in deciduous trees, below the canopy.

Two to four eggs are laid and are incubated for four weeks, the young fledging after about four to five weeks.

DISTRIBUTION AND HABITAT: An uncommon migrant in California, this hawk is basically an eastern species, breeding in central and southeastern Canada and the eastern half of the United States, and wintering in southern Mexico and northwestern South America. The individuals moving through California likely come from the westernmost limits of its breeding distribution. They enter California virtually unnoticed and, except at the Marin Headlands, hardly anybody ever sees them later; they appear there annually numbering from a handful to over 200 (Fish 2001).

In their eastern nesting ranges, Broad-winged Hawks inhabit chiefly moist deciduous woodlands.

SIMILAR SPECIES: This species is occasionally confused with the Red-shouldered Hawk *(Buteo lineatus)*. A perched juvenile California Redshoulder can strongly resemble an adult Broadwing, but note differences in tail length and tail bands and its longer legs.

REMARKS: Broadwings are so infrequent in our state that sighting one is a major event for the hawk-watcher.

SWAINSON'S HAWK   *Buteo swainsoni*
(Pls. 5, 8; Figs. 33, 94)

IDENTIFICATION: A large buteo that can look fluffy and rotund at times but usually appears more slender and streamlined than the similar-sized but heavier Red-tailed Hawk; overall, this species may be described as Redtail lite, appearing buoyant in flight and distinctly preferring small prey. The long wings enhance its attenuated appearance, the wingtips nearly or quite reaching the end of the tail, sometimes even beyond. A Swainson's Hawk often gives a mild-mannered impression.

Adults are highly variable, ranging in color from nearly entirely white below and brown above to dark rufous to black except for the pale undertail coverts. Some forms (morphs) are distinguished in this continuum.

The light morph has head and back parts dark brown, with a white forehead and a white throat, below which there is a dark brown or dark rufous bib sharply set off from the white belly. The underwing linings are white, contrasting with the dark gray brown remiges. It often has some rufous barring on the belly or flanks, sometimes rufous barring over the entire white underparts.

The dark morph is overall chocolate to black except for a white spot in front of the eye and the pale color of barred undertail coverts. The underwing linings are dark rufous to black brown, darker than the remiges. This morph has a conspicuous, broad, dark subterminal tail band.

The rufous morph has lower parts all rufous except for a white throat, without a dark bib, and back parts all dark brown. Alternatively, it may have a dark brown back, head, and bib, set off from the dark rufous, often heavily barred belly. The undertail coverts are white, occasionally with small spots. Underwing linings are white or rufous or both.

In juveniles, the back is mottled brown and beige, rather like that of a juvenile Red-tailed Hawk, but the forehead is pale, and the light cheek contrasts with a dark malar patch or stripe below the angle of the gape. The breast is cream, framed by dark brown, and the cream flanks are heavily spotted; in juveniles of dark morphs, the breast is heavily spotted as well. The beige leg feathers can be immaculate, or narrowly or broadly barred.

FLIGHT: While this buteo is soaring, the wings form a strong dihedral. The wingbeats are soft. The wings appear long, tapered, and distinctly pointed. The tail is short to medium, with many bars, the subterminal bar the widest. In juveniles, the dark remiges contrast with the beige underwing linings, which may be nearly immaculate or sparsely or heavily barred and spotted. This is the only buteo with all underremiges dark in all forms.

VOICE: A rather silent raptor, it rarely vocalizes except during the nesting season, when alarm calls are directed at humans and other predators; these calls resemble slurred, descending whistles. The young give begging calls.

FEEDING: Swainson's Hawks soar sometimes at great heights searching for food; they can catch flying insects on the wing, though most of their prey is terrestrial. Lower soaring, hovering, and kiting are all used when hunting rodents. Still-hunting is a common foraging method. Speculative low forays over prey-rich areas are also employed. Walking about on the ground is common and is the foraging method of choice when pursuing grasshoppers. Sometimes, dozens of Swainson's Hawks, chiefly juveniles, congregate in a field in quest of these insects, like a large flock of chickens, each hawk capable of eating more than 100 grasshoppers a day. Mammals up to the size of a young rabbit are taken, though more commonly voles, gophers, and small ground squirrels; amphibians and reptiles are occasionally caught, very rarely birds. Often, the main prey changes with the season and availability; predation is frequently tied to agricultural activities that make various prey species more vulnerable.

REPRODUCTION: Courtship includes both pair members soaring together near the nest tree and an undulating flight by the male, consisting of a short series of dives alternating with pull-ups in which the hawk allows its momentum to carry it upward again.

Given the choice, this hawk much prefers very tall trees for nest sites, such as old valley oaks *(Quercus lobata)* and cottonwoods in riparian settings (England et al. 1995); however, nests are often built in surprisingly small trees such as junipers, for lack of taller ones (Woodbridge, Finley, and Bloom 1995). In recent years, pairs have set up housekeeping in street and backyard trees in a few towns of the Central Valley, indicating a dearth of nest sites in nearby rich foraging areas—generally within 5 km (3 mi); some birds may even nest in bushes no taller than 2 m (6 ft) in such areas. Old nests of other species are sometimes used. As with many other raptors, the nest is garnished with fresh greenery, which may serve to keep away arthropod parasites and scavengers. Swainson's Hawk nests can be surprisingly small, much smaller

Fig. 94. A Swainson's Hawk on her nest in a cottonwood tree.

than those of Red-tailed Hawks *(Buteo jamaicensis)*, and the incubating female's tail and wings may project far over the edge.

Incubation takes about five weeks, the young fledging from the nest after six weeks or so, although they may start venturing out of the nest a week earlier. The youngsters depend on their parents for food for at least another month, although they soon set out after grasshoppers.

DISTRIBUTION AND HABITAT: The main breeding range of the Swainson's Hawk is throughout the Great Plains region extending barely into northern and northeastern Mexico. The California population consists currently of fragmented subpopulations.

Until the mid-1970s, Swainson's Hawks could be seen nearly throughout the state, at least in migration, with the exception of the extreme northwest and northern coast. Flocks of hundreds of migrants were commonly encountered in the great valleys and over our southern deserts. Although aggregations of 200 or so birds can still be observed on occasion, nesting populations throughout California have declined dramatically (by nearly 100 percent in southern California).

The Central Valley supports the greatest number of breeding Swainson's Hawks in the state. Other clusters of breeding birds can be found in Butte Valley and the Klamath Basin in appropriate habitat and in the northeastern corner of the state; small numbers may breed in Mono and Inyo Counties, a recent range expansion probably related to expanded alfalfa cultivation. In all of southern California, only three breeding pairs remain in the Antelope Valley (P.H. Bloom, pers. comm. 2003).

The great majority of Swainson's Hawks that nest in the Central Valley spend the winter in northwestern Mexico. These hawks avoid the hazards of extreme long-distance migrations that face other populations of Swainson's Hawks, including those of northeastern California, which, like others of the Great Basin, chiefly winter in Argentina. Most Swainson's Hawks arrive in the state in March or April and leave again in early September. Those in the Central Valley form communal groups in August following breeding, but the Great Basin populations tend to stay on their breeding grounds before migrating. Juveniles of this species sometimes do not migrate during their first winter. The 20 to 30 birds wintering in the mid–Central Valley may be migrants from elsewhere, not local breeders or their offspring (D. Anderson, pers. comm. 2004).

Swainson's Hawks are open-country buteos and are seen foraging in grasslands and agricultural areas. Often very tame, they perch low along roadsides on fence posts, telephone poles, and even on wires, hay bales, and low hummocks. Besides riparian woodland, nesting habitat includes piñon-juniper woodland, sagebrush, and savanna. Formerly, they were widespread nesters in Joshua tree *(Yucca brevifolia)* woodland in the Mojave Desert.

You have a very good chance of finding a Swainson's in late spring and in summer in the agricultural areas between Sacramento, Davis, Vacaville, and Stockton (such as at the Nature Conservancy's Cosumnes River Preserve), the Sacramento–San Joaquin Delta, and in agricultural valleys of the northeastern corner and the north-central part of the state (for example, in Surprise Valley, Modoc County; Big Valley and Honey Lake, Lassen County; and Butte Valley, Siskiyou County). Migrants are seen in southern California.

Although dark morphs are generally rare, they are more numerous in northern California than in other parts of this hawk's range, comprising 35 percent of the population (Brown 1996).

SIMILAR SPECIES: When seen in flight, all other buteos, including dark forms, have pale underremiges. The lighter top of the head of a dark morph Swainson's might be confused for the nape of a Golden Eagle *(Aquila chrysaetos)*, but an eagle is much larger, has feathered tarsi, and the wingtips do not reach the tailtip. A Northern Harrier *(Circus cyaneus)* female and juvenile are longer tailed, lack pointed wings, and have a much bigger white rump.

STATUS: Listed as threatened in California. Populations in other states are more stable, but the Swainson's Hawk is a federal Bird of Conservation Concern.

REMARKS: When egg collecting was fashionable and legal, small boys locally called this species the "Five Dollar Hawk," from the catalog price of its eggs (Sharp 1902).

ZONE-TAILED HAWK   *Buteo albonotatus*
(Pls. 4, 10)   `RARE`

IDENTIFICATION: A large, dark buteo appearing somewhat less plump and smaller than a Red-tailed Hawk. The bright yellow cere, gape, and legs are prominent. Wingtips of a perched bird reach the tip of the long tail or beyond. The legs look short.

The adult appears all black at a distance, but close up it looks dark gray or charcoal both above and below, the gray resulting from a bloom on the black feathers. The male has one wide and one narrow white band in the black tail; the female has one wide and two narrow white bands. When the tail is folded, the bands between the black bands (the interstices) appear gray from above, white from below. In flight, the undersides of the remiges are dusky and show many narrow bars.

The juvenile is brown black overall, with many narrow bands in the dusky tail, like the wings. The body often has few to many scattered, small, white spots that may be difficult to see.

FLIGHT: A Zone-tailed Hawk has long, narrow wings and a long tail, in outline much like a Turkey Vulture *(Cathartes aura)*. It flies with wings in a pronounced dihedral and in fact appears to mimic a Turkey Vulture, tilting like the latter from side to side.

VOICE: This species is more vocal than the Redtail *(Buteo jamaicensis)*. The call has been described as a catlike mewing, lower in pitch than the scream of its relative.

FEEDING: Unique among California buteos, this hawk forages almost exclusively from the air, gliding over prey-holding country at altitudes of 15 to 100 m (50 to 300 ft). It frequently flies in the company of Turkey Vultures and apparently uses its resemblance to them to get close to unsuspecting prey, at which it stoops from the air, usually at a shallow angle, using intervening cover to conceal its approach. Evidence is strong that this hawk's resemblance to the Turkey Vulture is indeed an example of aggressive mimicry rather than merely one of similar design for similar aerodynamic needs, as is occasionally suggested.

Mammals such as ground squirrels, birds (to quail size), and reptiles (especially lizards) are typical quarry.

REPRODUCTION: A series of long, steep dives alternating with pull-ups are part of the courtship display. Conifers and tall trees of riparian woodlands, such as cottonwood and sycamore, are most frequently chosen as nest trees.

From one to three eggs are laid, two being the most common; they are incubated for five weeks, and the young fledge in about seven weeks.

DISTRIBUTION AND HABITAT: With scattered populations in Arizona, New Mexico, and Texas, this species is chiefly found from Mexico southward through Central America and parts of South America.

Zone-tailed Hawks are rare in California. Being essentially

spillovers and winter visitors from Arizona and Mexico, they are chiefly seen in Orange, San Diego, Riverside, San Bernardino, Los Angeles, Ventura, and Santa Barbara counties, roughly from November to April. One sighting was recently reported in central California. Very small numbers probably breed in southern California with some regularity; as many as three pairs may nest in San Diego County.

Zonetails are sometimes seen in eucalyptus trees near the entrance to the San Diego Wild Animal Park (roosting there with Turkey Vultures) and are reported fairly regularly in winter along Wheeler Canyon Road and Aliso Canyon Road northeast of Ventura. Others have been seen flying over the city of Santa Barbara. Some individuals are thought to return yearly to the same California wintering areas (McCaskie and San Miguel 1999).

This hawk prefers two distinct types of habitat: riparian woodland (especially if backed by cliffs or rimrock) and rocky, rugged country with mountain coniferous forest and chaparral.

SIMILAR SPECIES: A Turkey Vulture's head appears very small by contrast and lacks the Zonetail's yellow cere and gape. When perched, a Common Black Hawk *(Buteogallus anthracinus)* is much longer legged and has a much bigger beak, and in flight, it is not at all shaped like a Turkey Vulture.

REMARKS: The association of this hawk with Turkey Vultures is so common that it pays to examine cruising flocks of these scavengers in likely habitat.

RED-TAILED HAWK   *Buteo jamaicensis*
(Pls. 6, 8; Figs. 9, 13, 17, 20, 23–27, 29, 31, 34, 56, 95)

Like some other prairie-savanna buteos, a very common hawk that is highly variable in both juvenal and adult feathering. Five races occur in the United States, two of which are found regularly in California, the most frequently seen form being the light morph of the race *calurus,* common in the western United States. The other race that shows up in California is the Harlan's Hawk *(B. jamaicencis harlani).* However, colors and patterns may vary widely between individuals in a given area within each race, and pairs of different morphs are common.

The most readily separable forms are described below, but there are all sorts of gradations between them, sometimes presenting an identification challenge; what is described as a light morph in the west has light undersides that can vary from pale tan to buff to white. Pure white Red-tailed Hawks are seen on occasion as well as partial albinos and leucistic birds, which are a very pale rufous color. Among Harlan's Hawks, there is a charcoal black dark form; other adults have variable white breast streaks on their otherwise charcoal black undersides, and still others are mostly white below. Generally, Harlan's are blackish where other Red-tailed Hawks are brown. Eye color ranges from pale straw yellow to deep reddish brown, the latter usually in adults; in Harlan's Hawks, the pale color may persist into adulthood.

IDENTIFICATION: A large, brown-mottled, big-headed hawk, often appearing broad waisted. Wingtips of a perched bird nearly or fully reach the tip of the tail in adults but are short of the tailtip in juveniles (which have longer tails). All ages and both sexes have a dark patagial patch on the spread underwing, easily seen in overhead flight. The feet and cere are yellow. Sexes in all forms are alike except for size, the female noticeably larger.

An adult light morph *calurus* has a head that is brown to golden ocher, with a large brown malar patch under the eye and a dark or streaked throat. The back, folded wings, and scapulars are brown with whitish or cream spots and patches that often form a rough V or U shape when seen from a distance. The underside is cream to pale rufous tan, often with a "vest" of dark streaks or spots that may be barely noticeable or very pronounced. The flank, belly, and leg feathers have rufous bars. The tail is rufous with a dark subterminal band and, sometimes, additional narrower dark bands.

An adult dark morph *calurus* is overall jet black or, more commonly, dark brown, appearing black from a distance. The tail is heavily barred, with a very wide subterminal band. Viewed in flight overhead, the pale remiges contrast with the very dark wing linings and are conspicuously barred.

An adult rufous morph *calurus* has a richly colored, dark rufous breast bordered by the very dark vest, head, and neck. The wing linings are dark rufous, the patagial patch darker yet. Undertail coverts are dark rufous with pronounced dark bars, and the barred dark rufous tail has a broad subterminal band.

An adult Harlan's Hawk lacks a red tail, almost entirely lacks rufous coloration in its various morphs, and it is blackish where other Red-tailed Hawks are brown. A charcoal black dark form exists. Other adults have variable white breast streaks on their otherwise charcoal black undersides. Still others are mostly white below. All, however, have tails that are marbled gray (occasionally also with traces of red). The tail is sometimes faintly barred.

Except for the darker forms, a juvenile Red-tailed Hawk is normally easy to identify. The basic pattern of the underparts (pale bib and dark vest) and barring of the tail are readily seen (the pattern may be obscured by additional pigment in the darker forms). The juvenile appears generally mottled brown and white or cream. The usually dark brown head has paler brow ridges, and there is a dark malar patch below the eye. The throat is all dark or heavily streaked. The folded wings and scapulars are mottled, with the nearly solid brown back forming a dark triangle; the brown of the back ranges from tan to chocolate. Most commonly, the vest of dark and light checkered feathers spanning the upper belly is pronounced, contrasting with the white to pale ocher bib on the breast and a pale lower belly. Leg feathers are closely barred. The tail bears numerous narrow dark bars, the subterminal one being the widest. Seen from above in flight, the wing is distinctly two toned, with primaries and primary coverts lighter than the secondaries and secondary coverts. In the dark and rufous morphs as well as in Harlan's juveniles, the bib may be partly or completely obscured by heavy streaks.

FLIGHT: This hawk often soars in circles, with wings arching somewhat forward and raised a bit above the horizontal, the tail fanned. Wing position is variable, but typically the wings appear broad and long and end in pinions that resemble the spread fingers of a hand. The tail is short to medium. In all juveniles, the wing linings are usually heavily marked, the patagial patch is pronounced, and the tail is noticeably longer than that of adults. Obvious barring of the remiges and conspicuous white "windows" at the base of the outer primaries are prominent features of the Red-tailed Hawk in flight but are not confined to that species.

VOICE: A Red-tailed Hawk expresses its displeasure with a slurring call that sounds so distinctly hawkish that moviemakers love to use it as a background sound to indicate a wild and lonely place. Sometimes it responds in this fashion to a human coming near even during nonnesting seasons, and the same call, or one very like it, is also directed at a Redtail owning an adjacent territory.

Fig. 95. Red-tailed Hawk nest in cliff.

Yelping calls, rather similar to those of the Golden Eagle *(Aquila chrysaetos)*, appear related to courtship, and the young produce begging calls. Because of this species' willingness to nest close to human activity, and because of their frequent use in movies and television commercials, some of these vocalizations have become very familiar to a great many people.

FEEDING: The Redtail forages in a variety of ways. In the west, the most common methods appear to be soaring, kiting, and hovering over prey-rich land, particularly over hilly terrain. It also still-hunts from perches (apparently the preferred method in the eastern United States) and uses speculative search flights. This raptor stoops like a falcon at times from great heights at flying prey; in California, Band-tailed Pigeons *(Columba fasciata)*, especially if weakened by trichomoniasis, are taken this way. Grasshoppers may be caught in foot chases, and carrion is not disdained. One adult male was observed hovering above a Badger *(Taxidea taxus)* digging out a ground squirrel burrow, apparently waiting for the rodent to bolt.

Most commonly, small to midsized rodents are the main food; however, amphibians and reptiles as well as rabbits and birds as large as pheasants are captured. Young ground squirrels are a frequent fare for the hawk's young, being available in numbers at just the right time. Some individuals become skillful hunters of city pigeons. Large quarry is plucked and eaten on the ground, whereas mice and such are taken to fence posts or limbs and, more often than not, eaten whole or in two bites.

REPRODUCTION: Some courtship displays also serve as territorial advertisement. Diving from great heights at speed and pulling out and upward is one such act, performed by the male. A common display by both pair members is high circling, with the legs hanging, and calling; often, the male, flying above the female, slowly swoops down and touches his mate's back with his dangling feet, or comes close to doing so. Pairs occupying adjacent territories meet at the boundary and, each pair circling on its home side, hurl invective at each other. Redtail territories cover a great range of habitats but always include several suitable nest sites and foraging grounds.

The stick nests, which can attain great size, are built in whatever large trees are available, commonly in eucalyptus, oaks, and sycamores, but also in conifers, aspens, and ornamental yard trees. Where they are available, the Redtail seems to prefer trees growing on slopes to flatland trees. It uses cliff ledges as well, and even human structures such as buildings and high-tension towers. When live oak (which offers much support) is the nest tree, a pair may use the nest for many years, adding to it yearly until it becomes enormous— eagle sized, though usually not as wide, and built of smaller twigs. In some parts of California, nests of different pairs are less than half a mile apart, indicating a generous food supply and an abundance of nest sites.

A Redtail typically lays three eggs, though some nests may hold one, two, or four. Incubation lasts about four and a half weeks, and the young remain in the nest for six and a half weeks.

DISTRIBUTION AND HABITAT: The Red-tailed Hawk is resident roughly from the Arctic Circle southward through all of Canada and the United States and most of Mexico and parts of Central America. It is ubiquitous in California from fall until the end of winter, and very nearly so for the rest of the year, even after the migrant individuals have left. This species has even been seen in mid-January in the snow-covered Sierra Nevada. Juvenile Red-tailed Hawks are the most often encountered large, brown-mottled hawk in winter in most parts of California.

Of all the buteos, the Redtail has made the greatest use of

human-made environments. It is the most commonly seen large hawk foraging over cultivated and pasture fields, airports, highway borders and medians, suburban play fields, and even cities. The hawks sit on exposed perches such as snags, fence posts, telephone poles, and light standards.

The Redtail breeds in valleys and foothills, in cities, over busy roads, and by remote mountain meadows. An adequate nearby food supply and suitable nest sites are all that are required, with the raptor willing to travel a half mile or more from its nest to foraging areas. Because this hawk is adapted for open-country hunting, fields, sage flats, marshes, or other open ground near the nest site are some of the most important components of nesting habitat. Young birds may disperse northward and eastward, at times moving as far as Idaho (Bloom 1985; Scheuermann 1996).

Harlan's Hawk, a regular winter visitor in California, is seen more dependably in some areas than in others (in the Central Valley, for example), especially in the northern and central parts of the state. It breeds in Canada and Alaska, and its main wintering grounds are in the central United States.

SIMILAR SPECIES: The Swainson's Hawk *(Buteo swainsoni)*, Ferruginous Hawk *(B. regalis)*, Rough-legged Hawk *(B. lagopus)*, and Broad-winged Hawk *(B. platypterus)* all have distinctive underwing and underside patterns.

REMARKS: Few raptors manage to look as serene, plump and harmless as a Red-tailed Hawk at rest, but it is a potent predator, and some fiercely attack humans that come too close to their nests. Adult resident Redtails often perch together, even out of the breeding season. Fledged young from different nests may band together in late summer and practice flying and foraging skills. This is also the time many first test their strength; in a suburban neighborhood, an ambitious juvenile female, aided by a slope and a strong wind, carried aloft a large, wildly flailing housecat, which survived with only minor injuries after the hawk dropped it from an altitude of about 15 m (50 ft).

In rural areas, this is the raptor that to this day is frequently called the "chicken hawk." Although Red-tailed Hawks rarely killed chickens in the days when these were mostly free ranging (most preferring gophers and ground squirrels and such), it was common to find the feet of many dozens of shot Redtails nailed to barn doors as late as the 1950s.

PLATES

PLATE 1   Vultures

See pl. 10 for additional flight images.

Turkey Vulture PAGE 157

Wings are frequently held in dihedral position as shown.

Black Vulture PAGE 156

Soars on flat wings.

California Condor PAGE 160

PLATE 1   Vultures

juvenile   adult

juvenile

adult

Turkey Vulture

Black Vulture

California Condor

adult

juvenile

adult sunning

PLATE 2   Kites, Osprey, Northern Harrier

White-tailed Kite
PAGE 169

On adult in flight, note black patch on underwing at the hand.

Mississippi Kite
PAGE 172

Falcon shaped in flight, but note short outermost primary.

Osprey
PAGE 164

On female in flight, note "necklace"; males may also have this feature. Note dark patches on underhand.

Northern Harrier
PAGE 174

Note facial ruff and white rump. On juvenile, rust color on underparts fades with age, and streaking may be absent. On adult female in flight, note conspicuous wing and tail barring; on adult male in flight, note trailing edge of wing and dark primary tips.

PLATE 2   Kites, Osprey, Northern Harrier

adult

juveniles

hovering

White-tailed Kite

adult

adult male

adult

juvenile

Mississippi Kite

juvenile

juvenile

adult

Osprey

female

male diving at prey

adult

juvenile

adult male

adult males

adult females

Northern Harrier

PLATE 3   Accipiters

Note long tails and rounded wings. Sexes look alike, except males are smaller.

Sharp-shinned Hawk

PAGE 178

Tailtip square or sometimes rounded (note juveniles). Long, thin (toothpick) legs. In flight, the short wings often appear "hunched" forward (note juvenile in flight).

Cooper's Hawk

PAGE 182

Tail very long, always rounded. In a soar, the longish wings are often nearly straight across their leading edges, with head and neck projecting far forward.

Northern Goshawk

PAGE 186

Tail long and rounded, legs short. On adult female, note striking black and white head. Adults and juveniles have conspicuous whitish superciliary. On adult in flight, note strongly barred primaries, nearly plain secondaries.

PLATE 3   Accipiters

juvenile female

adult male

juvenile female

juvenile female

adult male

juvenile female

juvenile

adult

Sharp-shinned Hawk

Cooper's Hawk

juvenile

adult

adult female

light juvenile male

dark juvenile female

juvenile

Northern Goshawk

adult

PLATE 4   Southern Rarities

Common Black Hawk
PAGE 191

On adult, note yellow cere and lips (gape) and legs. In flight, appears to have extremely broad wings. On juvenile, malar patch does not reach eye, superciliary is conspicuous, and tail bands are numerous and wavy.

Crested Caracara
PAGE 231

Juvenile like adult but browner. On adult in flight, note whitish panels in outer primaries, long neck and tail.

Zone-tailed Hawk
PAGE 205

Yellow cere, gape, and legs. Sexes differ in size and number of tail bands. On juvenile, variable numbers of white spots are often difficult to see and may be nearly absent. On adult in flight, note the dihedral position of wings and overall shape like a Turkey Vulture's (see also pl. 10).

Harris's Hawk
PAGE 192

Cere and face yellow. On juvenile, dark streaking of underparts is variable; streaking may nearly obscure pale ground color. On adult in flight, note unique rufous wing linings, extremely broad black band on white tail, and white upper- and lower lower-tail coverts.

PLATE 4 Southern Rarities

adult

juvenile

adult

Crested Caracara

adults

Common Black Hawk

Zone-tailed Hawk

adults

juvenile

juvenile

adult

Harris's Hawk

adult

juvenile

PLATE 5   Smaller Buteos and Swainson's Hawk

Red-shouldered Hawk

PAGE 195

Adult is brightly colored, very rufous with black-and-white checkered patterns. On juvenile, upperparts are speckled, underparts
both streaked and barred. On adult in flight, note striking black
and white barring of wings and tail. Note pale or white crescent in
hand of wing of all ages, visible from above and below.

Broad-winged Hawk

PAGE 200

Note very small size. Adult has broad tail bars. Juvenile has variably streaked underparts and tail with numerous narrow bands.
On adult in flight, note dark breast and wide, dark trailing wing
edges (less obvious in juveniles) in contrast with the generally pale
underparts; wings are short and wide but pointed in all ages.

Swainson's Hawk

PAGE 201

See pl. 8 for flight images.
Adult has several morphs (highly variable), with numerous intermediates. Light juvenile shown at two stages, with large dark
patches on sides of neck and upper breast (left) and, at end of first
year, after feather wear and bleaching by sun (right). Rufous adult
may lack dark bib and most barring shown here, instead sport
brick red underparts.

PLATE 5   Smaller Buteos and Swainson's Hawk

adults

juveniles

Red-shouldered Hawk

adults

juvenile

juvenile

dark adult

Broad-winged Hawk

light juveniles

dark juvenile

rufous adult

light adult

dark adult

Swainson's Hawk

PLATE 6 Red-tailed Hawk

Red-tailed Hawk

PAGE 207

See pl. 8 for flight images.

Light morph adult has mottled back and red tail; on light morph juvenile, note dark triangle on back, pale bib, and conspicuous dark vest. On rufous morph adult, breast is rufous to deep chestnut and tail bands are numerous; on rufous morph juvenile, bib is streaked, sometimes heavily so.

On dark morph adult, tail is heavily banded; dark morph juvenile sometimes has suggestions of dark rufous streaking on underparts. Adult intergrade shown is between rufous and dark morphs.

Harlan's Hawk adult is variable; marbled tail may show bars and traces of red, but appears whitish from a distance.

Adult resident Redtails often perch together, as does the mixed pair of light and rufous morphs shown here.

PLATE 6   Red-tailed Hawk

light adults

rufous juvenile

light juveniles

rufous adult

adult intergrade

dark juvenile

dark adult

rufous

light

Harlan's juvenile

Harlan's adult

Red-tailed Hawk

PLATE 7   Large Buteos

Both of these buteos are "booted," having feathered tarsi, with only the toes bare. See pl. 8 for flight images.

Rough-legged Hawk PAGE 216

Often fluffy looking, with a big, round head and small beak and feet.

On light morph adult male, note extensive mottling, especially on underparts except for lower breast. Light morph adult female has black brown cummerbund broken along midline of belly and single black tail band. Light morph juvenile has solid black brown cummerbund across belly; terminal half of tail is dusky but lacks distinct band.

On dark morph adult female, head is often paler than body, with single wide subterminal tail band. On dark morph juvenile, head is sometimes as dark as body; terminal half of tail is always dusky.

Ferruginous Hawk PAGE 213

Very large. Busty chest; big flat head.

Light morph adult is colorful but may appear darker in poor light; underparts are nearly pure white to rufous with conspicuous barring. Light morph juvenile has dark line through eye. Dark morph birds range from rufous (shown) to nearly black. On dark morph adult, note paler undertail coverts; on dark morph juvenile, head and breast contrast with darker body. Note pale tail bars.

PLATE 7 Large Buteos

light adult male

light adult female

light juvenile

dark juvenile

light juvenile hovering

Rough-legged Hawk

dark adult female

light juvenile

light adults

dark (rufous) juvenile

dark (rufous) adult

Ferruginous Hawk

PLATE 8 Prairie-savanna Buteos in Flight

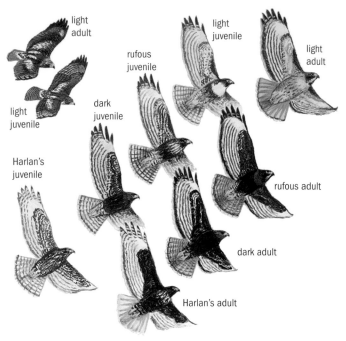

Red-tailed Hawk has dark patagial patch on underwing (obscured in dark morphs). All hawks shown are calurus except Harlan's. On light morph calurus juvenile, two-toned wing is apparent from above.

Ferruginous Hawk has rather pointed wings, often mostly whitish underparts. On light morph adult, leg feathers form a dark V, and reddish tail with white base is apparent from above; dark morph adult and dark morph juvenile are very similar.

PLATE 8 **Prairie-savanna Buteos in Flight**

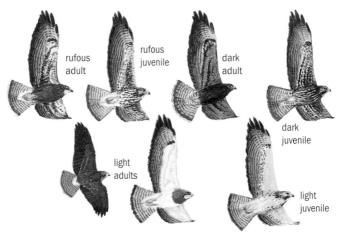

Swainson's Hawk has fairly pointed wings, dark underremiges. On light morph adult, note light underwing coverts and conspicuous white on upper tail coverts (apparent from above).

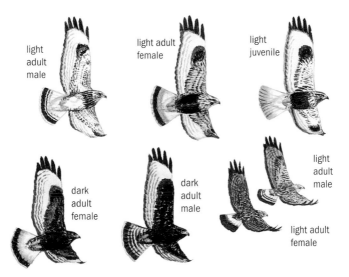

Rough-legged Hawk has fairly parallel-edged wings with blunt tips. On light morph juvenile, note black patch on underhand.

PLATE 9 **Eagles**

Both species are very large. See pl. 10 for flight images.

Golden Eagle PAGE 224

Often appears long and somewhat slender when perched. Feathered tarsi; large, two-toned beak. Adult upperwings are mottled, banded, or both, from retention of older, bleached feathers. On juvenile, basal half of tail gleaming white. Subadult has slight mottling on upperwings; white basal half of tail looks dirty, especially on central tail feathers.

Bald Eagle PAGE 220

Appears squat and bulky when perched. Short, bare tarsi; very large beak. Juvenile's tail is mottled white and dark brown; beak is black. Second-year subadult has much white on underparts and dark beak. Third-year subadult is like second-year, but with whitish malar area; beak begins to turn yellow. White triangle on back of third-year bird may be present on second-year birds as well.

PLATE 9 Eagles

juvenile

adult

Golden Eagle

subadult

juvenile

subadult
(third year)

adult

Bald Eagle

subadult
(second year)

PLATE 10 Large, Dark Raptors in Flight

Zone-tailed Hawk has long narrow wings, long tail; resembles Turkey Vulture.

juvenile
(slow glide)

adult

adult
(fast glide)

Turkey Vulture has small head, silvery flight feathers from below; juvenile is shown flex-gliding at low speed (above) and adult at high speed (below).

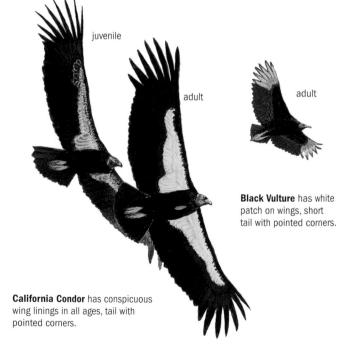

California Condor has conspicuous wing linings in all ages, tail with pointed corners.

Black Vulture has white patch on wings, short tail with pointed corners.

PLATE 10   Large, Dark Raptors in Flight

Golden Eagle viewed from above, has uniform brown or brown black coloring as juvenile, except for nape and tail zones (and wing patch, if present); as adult, shows buffy bars formed by bleached and unmolted upper wing coverts; as subadult, has combination adult/juvenile tail and beginnings of upper wing bars.

juveniles

subadult

adults

juvenile

adult

subadults (second year)

Bald Eagle has huge beak visible from afar; head and neck protrude more than half of tail length. Projecting secondaries of subadults are unmolted feathers from juvenal plumage.

PLATE 11 Small Falcons; Large Falcons in Flight

Merlin

PAGE 236

Broken, often indistinct malar stripes; underparts always streaked. Juvenile resembles adult female except in black race, in which juvenile male is usually darker than adult and juvenile female.

American Kestrel

PAGE 232

Often looks rotund. Large head with two conspicuous vertical bars on sides of head (one a malar stripe); pale underwings. Juvenile female resembles adult female.

Peregrine Falcon

PAGE 243

Anatum race shown. Note conspicuous face pattern, dark head, long, pointed wings, dark underwings, compact body.

Prairie Falcon

PAGE 249

Pointed wings shorter than Peregrine's; dark armpits.

Gyrfalcon

PAGE 241

PLATE 11 Small Falcons; Large Falcons in Flight

Merlin
(Prairie Race)

Merlin
(Taiga Race)

adult females

adult male

adult
male

adult
female

adult
male

adult female
(or juvenile)

adult male hovering

adult
female

juvenile
male

adult
female

adult
male

adult
male

female

female

Merlin
(Black Race)

American Kestrel

adult

juvenile
soaring

adult

gray
juvenile

Peregrine Falcon

Prairie Falcon

Gyrfalcon

PLATE 12   Large Falcons

See pl. 11 for flight images.

Peregrine Falcon
PAGE 243

All races and ages have pronounced malar stripe. All adults have barred underparts below breast.

Prairie Falcon
PAGE 249

Adult cheek often "dirty"; small round spots on underparts, largest on flanks. Juvenile cheek usually clean white to buffy; back brown; buffy or white underparts are narrowly streaked with drop-shaped marks, wider on the flanks.

Gyrfalcon
PAGE 240

PLATE 12 Large Falcons

Peregrine Falcon
(Tundra Race)

adult

juvenile

juvenile

adult

Peregrine
Falcon
(Anatum
Race)

adult

juvenile

Peregrine Falcon
(Peale's Race)

gray juvenile

gray adult

Prairie
Falcon

juvenile

adult

Gyrfalcon

FERRUGINOUS HAWK   *Buteo regalis*
(Pls. 7, 8)

IDENTIFICATION: A very large buteo with a busty chest and big, flat head. The wingtips of a perched bird reach nearly to the tailtip in the adult but are well short of the juvenile's tailtip. Most individuals have underparts that appear gleaming white from a distance. The enormous gape is visible from afar because it is bordered by thick, bright yellow "lips" that continue into the yellow cere. The legs are feathered to the toes.

The adult light morph's great size and colorful pattern make it easy to identify. The head is usually pale, and the underparts are white or cream. The leg feathers are variably dark rufous and finely barred. In some, the leg colors, and particularly the barring, extend to the flanks and belly, or the entire underparts may be pale rufous with heavy streaks and bars. The back, scapulars, and lesser upperwing coverts are rufous and broadly streaked with brown black, whereas the greater coverts and secondaries are gray with dark barring. The tail is pinkish, with a white base. In overhead flight, the color on the legs forms a very contrasting dark V; the wing linings have much white.

Adult dark morph color ranges from rufous or chestnut to blackish all over, or a combination of the two, visible only close up and in good light. Whitish breast streaks are variable or absent. The tail is pale gray above, as are the leading edges of the folded primaries. This morph is starkly two toned in overhead flight, the dark wing linings, head, and body contrasting sharply with the gleaming undersides of the flight feathers. Some authors separate rufous and dark morphs, but they are a continuum.

The juvenile light morph has a dark brown back. The uppertail is inconspicuously broadly barred. A dark line runs through the eye. Underparts are almost entirely white, although just-fledged young have buffy breasts that fade from sun exposure. The legs and flanks have a variable (but never heavy) scattering of spots.

The juvenile dark morph has a dark rufous head and breast, the rest of the body all blackish brown, or blackish brown entirely. Underparts are like the adult, except the tail is faintly barred.

FLIGHT: While soaring, the wings are held in a strong dihedral. Wingbeats are powerful, shallow, and fast. In flight, the chest looks conspicuously large and rounded, the whole body shaped rather like a football. The wings are long and taper to a fairly sharp point. When seen from above, the primaries display whitish bases that form "windows" brightly set off from the dark secondaries, as conspicuous as the pale base of the tail.

VOICE: A rather silent hawk, this buteo is most often heard when alarmed by a human presence near its nest, an event highly unlikely to occur in most of our state, because the Ferruginous Hawk rarely nests in California. Wintering individuals, at the sight of humans, very occasionally give the same alarm call, either a single drawn-out "keeah," or a wavering series of that sound.

FEEDING: This species is primarily a hunter of small to medium-sized mammals and normally catches only insignificant numbers of insects, amphibians, reptiles, and birds. Ground squirrels of various species are a common prey, and in some areas, jackrabbits or pocket gophers. Ferruginous Hawks avail themselves of a variety of foraging methods; still-hunting is frequently used, with telephone poles, fence posts, dirt mounds, and even flat ground serving as perches from which the hawk sallies forth or pounces. They also hunt from a soar and from a hover, and intending to surprise prey, with low level speculative flights at moderate speed. Perhaps the most spectacular method is a speculative flight that carries the hawk at high speed with wings angled back, close to the ground, over hundreds of yards. When prey is spotted, this long-distance dash terminates in a slanting dive downward, the raptor often tilting sharply over one wing.

REPRODUCTION: The heart of this hawk's nesting habitat is the Great Plains; it generally does not breed in California but has done so on occasion in the very north and northeast of the state (Modoc County and, more recently, Butte Valley in Siskiyou County [B. Woodbridge, pers. comm. 2004]). Besides the circling and calling together of the male and female, there appear to be no obvious courtship displays, and display flights such as flutter-gliding are mainly for announcing and defending territories.

Nests are placed in trees, telephone poles, platforms, and cliffs, on haystacks, and even on the ground; ground nests are often built on bluffs and are used for many years, the hawks adding more material annually. As a result, some nests become massive towers, reaching heights of over 4 m (12 ft). Oddly, nearly all nests contain dried cow pies (dung) in addition to more traditional building material. Nest defense can be vigorous.

The Ferruginous Hawk lays from three to four eggs, which are incubated for about four and a half weeks. The young fledge about five to six weeks after hatching. Numbers of active nests and young fledged vary with annual differences in prey abundance, with low-prey years resulting in up to a 50 percent reduction of young produced in a population.

DISTRIBUTION AND HABITAT: The Ferruginous Hawk breeds from south-central and western Canada through the Great Plains of the Great Basin to southern New Mexico and northern Arizona and northwestern Texas. It winters in the southwestern United States through Texas and southward to central Mexico.

In California, Ferruginous Hawks appear in early fall (usually mid-September) and can be found nearly throughout the state in areas of suitable habitat and prey until early April. Although they have been recorded in every county, in some they are exceedingly rare winter visitors; however, they can appear numerous in some years in the Great Basin country of the northeast and the grasslands of the state's interior valleys, and they can also be found in coastal valleys and parts of the Mojave Desert. They are seen in somewhat surprising numbers (for a steppe-grassland species) in river bottoms of the northwest (Hunting 1998). They are also found east of the Sierra Nevada. They perch commonly on the ground and are often quite fearless when perched on poles and high-tension towers, allowing close approach, especially in a car. In some areas, they aggregate in late afternoon and form communal night roosts, for example, on a high-tension tower.

This species appears highly specialized for hunting over open, sweeping land. Individuals trained in falconry fly poorly in enclosed country, for them an unnatural habitat. That does not mean, however, that the species cannot turn up in unexpected places. A large buteo seen from afar that was initially dismissed as just one more Red-tailed Hawk (*Buteo jamaicensis*) turned out in fact to be a Ferruginous when examined through binoculars; the raptor was perched on a snag on a wooded hillside of the inner Coast Ranges, next to a small meadow. Normally, however, wintering Ferruginous Hawks seek out grasslands, low hills, agricultural areas, shrub steppe of sage (*Artemesia tridentata*) and saltbush (*Atriplex* spp.), desert scrub with creosote bush (*Larrea tridentata*), desert, and piñon-juniper woodland.

The Carrizo Plain (San Luis Obispo County) is well known for its wintering Ferruginous Hawks, as is the Cuyama Valley (Santa Barbara County. In San Diego County they tend to congregate wherever there is grassland; many are seen at Lake Henshaw and the Ramona Grasslands. The Bureau of Land Management has identified its Harper Dry Lake (San Bernardino County) as a Key Raptor Area, in part because it includes an important communal roosting site for these hawks.

They can be seen in the Livermore Valley and Altamont Pass (Alameda County) in winter, and they can be found with near certainty on Burns Avenue and Burns Road just north of Bethany Reservoir in Alameda County. The southeast portion of Patterson Pass Road in Alameda County is also frequently rewarding, as are the southern San Joaquin Valley wildlife areas (Merced and Fresno Counties). Fewer are seen in the agricultural valleys of the southern San Mateo coast. In northern California they can be found in grassland and agricultural habitats of the Modoc Plateau.

SIMILAR SPECIES: Compared to a Red-tailed Hawk, a Ferruginous has faster and more flowing (not choppy) wingbeats with fewer glides, pointed wings, and a longer-looking tail. When the pointed wings are angled back in low-level, high-speed flight, this hawk may be mistaken for an enormous falcon (see pl. 7, Ferruginous Hawk in flight). The upperwing pattern of a juvenile Ferruginous Hawk resembles that of a juvenile Rough-legged Hawk *(Buteo lagopus),* but wing shape differs. The undertails of these juveniles also are similar, but the underwing patterns are very different.

STATUS: A California Species of Special Concern, and a federal Bird of Conservation Concern outside California.

REMARKS: The scientific name of this raptor translates as "royal hawk," an apt name for this stately, beautiful bird, the largest of the buteos. Ferruginous refers to the rusty color of the adult.

ROUGH-LEGGED HAWK   *Buteo lagopus*
(Pls. 7, 8)

Comes in a truly perplexing variety of patterns. Not only are there light and dark morphs and all manner of wretched intermediates between the two, but also, in the adults, males may differ from the females. The dark morph is very uncommon in the west.

IDENTIFICATION: A large buteo that looks often very fluffy, with a big round head and noticeably small beak for such a big hawk. The wingtips of a perched bird come almost to the tip of the tail or beyond. The tarsi are feathered to the small yellow feet.

The adult male light morph has a head finely streaked and appearing grayish, as do the back and wings, which are mottled black brown and gray, but it is never as sharply mottled white and brown as a Redtail *(Buteo jamaicensis)*. The bib is very broadly streaked with dark brown, producing a mottled appearance; the lower breast is white, with few or no spots. The belly is white or varyingly spotted, sometimes heavily. Underwing coverts have variable spotting; the entire wing has a black trailing edge. The lower third of the tail is conspicuously dark, the remainder white with variable (but always few) dark bands.

The adult female light morph has a buffy head. The brown back and wings are mottled with pale tan. The buffy bib is streaked variably, the lower breast is whitish or buff, the flanks are black brown. A very wide black brown cummerbund is divided along the midline. Buff, heavily marked underwing coverts contrast with mostly white flight feathers; there is a black trailing edge on the wings, and a single wide black band at the tip of the tail.

The adult dark morph appears black with a white forehead and a black-and-white barred tail, which frequently is not exposed. Underwing linings are all black, flight feathers often dusky with variable barring. The black parts of the male can be jet black.

The juvenile light morph has a pale or buffy or whitish head with variable fine streaks, a pale nape with streaks and a central dark triangle, dark brown back and wings, the feathers edged with buff or white. It has a white or buffy bib with variable streaks and has a very broad black brown cummerbund (which is undivided) across the belly. Underwing coverts are cream to buffy with variable spotting, and there is a conspicuous black patch on the underhand and a dusky trailing edge of the white, sparsely barred flight feathers. The tailtip is dusky and wide.

The dark morph juvenile is like an adult but may have a lighter head color.

FLIGHT: The wingbeat is shallow, soft, and fairly rapid, making the bird appear lightweight compared to a Redtail. In a soar, the wings form a shallow dihedral. In light morphs, prominent blackish patches on the undersides of the hand contrast sharply with the white outer primary bases. In all forms, a clear white area at the base of the outermost four or five primaries contrasts sharply with their blackish tips and the variable barring on the other flight feathers; the basal third of the tail is conspicuously paler than the dusky terminal third.

VOICE: This species is generally silent on its wintering grounds, although occasionally individuals may give the alarm call typically uttered near the nest: a descending, high-pitched, rough whistle.

FEEDING: A Rough-legged Hawk forages by still-hunting from telephone poles and other elevated perches but perhaps more commonly by kiting and hovering at no great height over rodent-rich fields. The flashing darks and lights of a hovering Roughleg are a real eye-catcher (see pl. 7, Rough-legged Hawk hovering). This species also forages by coursing low over promising land.

The Roughleg feeds chiefly on small rodents. Shrews are taken as well as some birds, especially while breeding, and even fish. Carrion, such as road-killed rabbits, attracts Roughlegs in winter, often leading to the hawks' demise under the wheels of vehicles.

REPRODUCTION: In North America, Rough-legged Hawks breed only in Alaska and in the Canadian Arctic. Courtship displays involve high circling of the pair and calling, and undulating flights may follow during which the male alternates between steep dives and sharp pull-ups to stalling speed only to dive once more. Nests are placed in trees where available, or else in or on cliffs and bluffs.

The number of eggs laid each year appears related to the number of available prey, which, in the Arctic, means lemmings; clutch size ranges from two to seven. Incubation lasts about four and a half weeks, the young fledging after another five or so weeks.

DISTRIBUTION AND HABITAT: This is a circumpolar species of the Holarctic; in North America it breeds only in the far north and winters in parts of southern Canada and continentwide in the United States, with the exception of the southeastern part and southern Texas.

While wintering in California, Rough-legged Hawks can be found throughout the state but are distinctly scarce in the southern counties. Their numbers are highly variable from year to year, and there are conspicuous good flight years. They usually show up in the north of the state with the onset of cold weather (commonly October or November), and most leave again in April.

Roughlegs seek out coastal and interior grasslands, sagebrush, agricultural lands, and other open country. They are numerous in the Klamath Basin and on the Modoc Plateau (Siskiyou and Modoc Counties); the Ash Creek Wildlife Area in southwestern Modoc County has many, and they occur in the sage steppe east of the Sierra Nevada. In those areas, they and Ferruginous Hawks *(Buteo regalis)* can be more common than Red-tailed Hawks (Small 1994). Roughlegs also show up in numbers along the northern coast grasslands and ranges and in the Central Valley. Fewer winter on the Carrizo Plain and in the Antelope Valley and at Lake Henshaw. Some appear in the Bay Area every winter, chiefly in the eastern Livermore Valley (Alameda County), and a single individual comes nearly every year, not only to the same area (in Coyote Hills Regional Park in Alameda County) but to the same snags. Surprisingly, it is not always the same bird.

SIMILAR SPECIES: A Red-tailed Hawk has relatively wider wings that appear shorter; a Harlan's Hawk has a marbled tail. A dark morph Ferruginous Hawk has more pointed wings than a dark morph juvenile Roughleg. A Swainson's Hawk *(Buteo swainsoni)* has very pointed wings. A perched adult male dark morph can be so black it may be confused with a Common Raven *(Corvus corax)* at a glance.

REMARKS: A Rough-legged Hawk, like a Gyrfalcon *(Falco rusticolus)*, has unusually dense plumage with more feathers per area of skin than does, for example, a Red-tailed Hawk; this accounts for the Roughleg's fluffed-pillow appearance.

Eagles

The term "eagle" is quite arbitrary; eagles have evolved several times from very different lines of hawks, and the two species found in California, the Bald Eagle *(Haliaeetus leucocephalus)* and the Golden Eagle *(Aquila chrysaetos)*, are not closely related, with the Bald Eagle showing anatomical similarities to kites and the Golden Eagle to buteos. Nor are eagles necessarily very large: one tropical Southeast Asian species is pigeon sized. The great size of our eagles, however, is one of their surest identifying traits.

In shape, our eagles resemble an oversized prairie-savanna buteo with extra long wings. The legs are short (though with a surprising reach in Golden Eagles), the toes thick and heavily armed for dealing with substantial prey. The large, deep beak is often conspicuous, particularly in Bald Eagles, much more so than that of a hawk. Unlike buteos, Golden Eagles never have pale bellies, but Bald Eagles do before they reach sexual maturity.

Both species soar frequently and glide over great distances on enormous wings.

BALD EAGLE   *Haliaeetus leucocephalus*
(Pls. 9, 10; Figs. 28, 96)

IDENTIFICATION: A very large eagle that, regardless of age, has a very large bill and a short tail. It appears bulky (broad shouldered and wide bodied). The wingtips of a perched bird do not reach the tip of the short tail. The feet and stubby legs are yellow. The sexes look alike, the male smaller.

The familiar adult pattern (dark body with white head and tail) is not fully developed until the fifth and sixth years. The adult's beak and eyes are yellow.

The juvenile, fresh out of the nest, is overall dark brown. It fades over the next several months to paler brown and tan. In good light, the breast is clearly darker than the belly, the beak is a dark black brown color, the cere blue gray, and the eyes are dark. The underwings, especially the coverts, have white mottling, and the sooty brown tail usually has some dirty white mottling. The upperwing coverts may have pale tips.

In the second year, the whitish or white- and dark-blotched belly is clearly set off from the dark brown breast, and there is a whitish or spotted triangle high on the back (the mantle). The mottled dark and white underwings persist, identifying this species even when the mottled tail has an indistinct dark terminal band reminiscent of a Golden Eagle's *(Aquila chrysaetos)* tail.

In the third year, a pale crown and throat are separated by a broad band running backward from the eye, and the beak begins to turn yellow along with the cere and eyes.

In the fourth year, the head and neck are white with dark streaks, the tail whitish except for dark edges and terminal band, and the mottling of the underwings is much reduced.

At long last, in the fifth and sixth years, the clean adult pattern is finally fully developed.

FLIGHT: A Bald Eagle soars on flat wings or wings slightly arched down. Flapping flight is with ponderous, slow wingbeats, more so than in the Golden Eagle. The wings are very long, the tail is noticeably short, and the head and neck project far forward.

VOICE: This species is noisier than the Golden Eagle, especially around conspecifics, when their chittering calls draw attention.

FEEDING: The Bald Eagle is decidedly more opportunistic in utilizing foods than the Golden Eagle; a Bald often robs other raptors and may even exploit garbage dumps. Carrion of all sorts is quite acceptable, including dead fish, and two juveniles in Canada were observed feeding on a partly submerged human body (Olsen 1986). Bald Eagles quickly learn to take advantage of the happy windfall presented by groggy fish turned loose by fishermen at catch-and-release fishing sites.

Although it is often looked down upon as a scavenger, the Bald Eagle can be in fact a highly adept and speedy hunter as well. It is quite capable of catching large waterfowl both in the water and in the air, even species as large as geese. Waterfowl, especially coots, usually constitute the main food of wintering Bald Eagles in California, whereas nongame fish make up the bulk of the diet during breeding season. Unlike the Golden Eagle, the Bald Eagle forages chiefly from perches, typically from snags near water and

Fig. 96. A subadult Bald Eagle (left) about to rob a fiercely mantling subadult Golden Eagle of its prey. Sometimes the tables are turned.

rarely from a soar. Very large fish and birds are towed to shore and eaten there, whereas smaller prey is taken to perches. The Bald Eagle sometimes "tests" rafts of waterfowl by swooping over them; scattering in alarm, the flock reveals sick or disabled individuals in its midst, which are then taken on a subsequent pass. Coots are sometimes forced to dive until too exhausted to do so and are then easily taken.

REPRODUCTION: Courtship in paired Bald Eagles in California begins in January and February. The pair bond between Bald Eagles appears to be permanent, and although fidelity to the nest site may be the actual cause keeping couples together, there is some evidence that migrating adult pairs stay together on the wintering grounds as well. Courtship displays are aerial and include pursuit flights, undulating flight (a type of display widespread in raptors), and pinwheeling.

The stick nests are nearly always built near water, although one was found several miles inland from Lake Cachuma near Santa Barbara (B. Mahall, pers. comm. 2003). In northern California, ponderosa pines *(Pinus ponderosa)* that project above the forest canopy are usually chosen; elsewhere, conifers are clearly preferred over other kinds of trees. More material is added yearly to the same nest, eventually resulting in a truly massive structure up to 6 m (20 ft) tall and nearly 3 m (9.5 ft) across at the top; there are often alternate nests as well. Bald Eagle nests are frequently in more exposed sites than are the nests of Golden Eagles.

Usually two eggs are laid, though at times just one, or even three. The young hatch after five weeks of incubation, carried out by both parents. Asynchronous hatching may result in siblicide. The young remain in the nest for about 12 weeks, and fledged young remain with or near their parents for another month or so before dispersing in late July.

DISTRIBUTION AND HABITAT: The Bald Eagle breeds across Canada and Alaska (except for the extreme north), down the eastern and western coasts into Mexico, in the Rocky Mountains, and locally elsewhere in the western states. It winters in midcoastal and central Mexico.

In California the Bald Eagle is nowhere common, except in the extreme north in winter; relatively large numbers are found along the Pit River (Shasta County) and associated reservoirs. The breeding population, reduced to 20 to 30 known pairs by 1974, is increasing and expanding, with breeding territories now in about half of California's 58 counties. Most breeding occurs in Shasta, Modoc, Siskiyou, Lassen, and other mountainous counties in the northern part of the state. The rest of the breeding population is more widely scattered, particularly in southern California. This eagle takes advantage of habitat created by reservoirs, particularly in southern California but also other areas, such as Del Valle Reservoir in Alameda County. A nest near Lake Hemet, a reservoir in Riverside County, was documented in 2003 as the first successful nest south of the Tehachapi Mountains in more than 70 years (R. Jurek, pers. comm. 2004).

Most of California's breeding Bald Eagles (excepting those at high elevations) remain in their nesting areas all year. Radio-tracked juveniles from northern California nesting areas moved northward, as far as southeast Alaska, to areas where stream-spawning fish arrive from the Pacific in late summer (Hunt et al. 1992). Some eventually return to the areas where they were born; others wander and potentially could expand Bald Eagle ranges (Jenkins et al. 1999).

Wintering eagles appear at lakes and reservoirs large and small nearly throughout California, sometimes in numbers, though never as many as are seen in the Klamath Basin National Wildlife Refuges shared with Oregon. Tule Lake is located here, famous for its tens of thousands of migrant waterfowl that attract hundreds of Bald Eagles, one of the largest U.S. wintering populations outside of Alaska. Smaller numbers spend the winter at the larger lakes in Lassen, Shasta, and Trinity counties and at Lake Almanor (Plumas County). In southern California, the greatest numbers are found at lakes such as Big Bear and Baldwin lakes in the San Bernardino Mountains.

Bald Eagles also winter at Lake San Antonio (Monterey County), Comanche Reservoir (Amador County), Lake Cachuma (Santa Barbara County), some of the larger reservoirs in the southern Sierra Nevada foothills, and on the Fall, Truckee, and Walker Rivers (Shasta, Nevada, and Mono counties).

Besides lakes and rivers, Bald Eagles are typically seen on islands, seashores, and lagoons, although occasionally they are encountered far from water such as in the agricultural Butte Valley (Siskiyou County). They seem especially fond of snags but also perch on high-tension towers, fence posts, and even on the ground.

Migrants wintering in northern California tend to arrive about November, sometimes a little later in southern California, and they leave around April. Small numbers are found between October and late March wintering at reservoirs in the San Francisco Bay Area, with some juveniles arriving as early as August.

SIMILAR SPECIES: The adult is unmistakable, but the juvenile and the subadult are often mistaken for a Golden Eagle. The Golden Eagle's tail looks long, and head and neck protrude only half the length of the tail in flight. A juvenile Bald Eagle's beak is larger and all dark (two toned in the Golden). A Golden Eagle juvenile's white and black zones on the tail and white windows on the underwings are clean, not mottled, and the white on the underwings is restricted to the bases of the primaries. The wings of a Golden Eagle are usually held at a slight dihedral or flat but not arched down. A Bald has a more flowing wingbeat than a Golden.

STATUS: Federally threatened but proposed for delisting. In California, it is listed as endangered.

REMARKS: A Bald Eagle may peacefully feed side by side on large carrion with a Golden Eagle, but it may bully the latter into relinquishing smaller prey; at other times, the tables are turned. Young Balds in their first and second years often parasitize Goldens.

GOLDEN EAGLE   *Aquila chrysaetos*
(Pls. 9, 10; Figs. 14, 19, 28, 32, 35, 96)

IDENTIFICATION: A very large brownish to blackish raptor with a bulky body and a conspicuous nape that ranges in color from nearly white to straw yellow to rufous orange. Wide across the shoulders when seen face-on or from behind, the rest of the body is of about equal width. The wingtips of a perched bird do not reach the tailtip. The head appears relatively small but not tiny. The beak is large. The legs are feathered to the bright yellow toes; the talons are black. The sexes differ in size only, the female noticeably larger. The female's head appears distinctly smaller (but with a somewhat larger beak) than the male's in proportion to body size, a trait that can be seen from a great distance, giving her body a more massive appearance.

In the adult, the upperwing coverts and scapulars molt irregularly, and older feathers from at least two previous years persist. Because of sun exposure, these fade from dark brown to tan to ocher and finally to pale straw and form conspicuous bands and patterns on the upperwings and scapulars. The adult occasionally has a white patch on the bend of the spread wing. In flight, the underwing is bicolored, the linings much darker than the grayish flight feathers, the latter with dark tips that form broad trailing edges on the wings. The tail is less distinctly barred from below but has a broad dark terminal band.

Younger birds appear solid dark brown except for the nape and the tail, which is conspicuously two toned. In the first four years, the plumage assumes adult coloring, especially noticeable in the tail. In the juvenile (first year), the basal half of the tail is white (prominent from afar), and the rest is black brown. The white wing patch at the base of the inner primaries, seen in flight, is always clean if present. In the following three years, the tail progressively becomes darker, with some white persisting closest to the body and on the lateral tail feathers. At the same time, bands appear on top of the wings formed by unmolted sun-bleached feathers. In the juvenile, there is often a white patch in the otherwise dark flight feathers at the arm/hand junction, variable in size, which normally disappears in successive molts.

FLIGHT: The wings are held in a moderate dihedral or sometimes flat, with shallow wing flexes rare while gliding in wind. In a soar, the outermost primaries are often splayed into conspicuous "fingers."

VOICE: The Golden Eagle is remarkably quiet as raptors go; vocalizations are not very loud and include a territorial call most commonly given at dusk during the breeding season, a rather similar greeting call, and the juvenile's begging call.

FEEDING: Like all raptors, the Golden Eagle is an opportunist and takes what is most easily available, including carrion, such as a dead cow or a road-killed deer, though other live food may be plentiful. It takes reptiles and even fish, although most commonly the prey is mammals, followed by birds. Rabbits and hares rank high as food, and in much of California, ground squirrels of various species form the staple, along with occasional fawns. The remains of badgers, bobcats, housecats, and foxes turn up in nests, and in the Great Basin, two or three eagles may combine their efforts to take down a grown deer in deep snow. Sometimes they hunt lesser quarry, such as jackrabbits, in pairs. In winter, when other food is less available, many Golden Eagles turn their attention to large birds, such as waterfowl, but they also commonly eat herons and Red-tailed Hawks *(Buteo jamaicensis)* and, occasionally, Turkey Vultures *(Cathartes aura).* Smaller birds are sometimes brought to the nest, such as magpies, flickers, and even birds as small as starlings. Clearly, the Golden Eagle is a versatile predator, though its hunting skills may take a year or more to fully develop; first-year birds often survive on carrion or they appropriate the prey of others. Although Golden Eagles do occasionally kill livestock, in the great majority of cases where eagles have been observed feeding on lambs, it has been demonstrated that the lambs in fact were killed by something else, and the eagles were simply feeding on carrion (skinning the lamb quickly reveals the presence of hemorrhaging puncture wounds that result from an eagle attack).

This eagle most commonly uses two foraging methods. Flying leisurely at altitudes of 45 to 900 m (150 to 3,000 ft) and diving on quarry is one; the other, more common, is contour-hunting: power-flying and gliding at speed two or three feet from the ground and following the lay of the land in hopes of surprising prey over a rise. Sometimes, the male and female of a pair hunt together, with one flying high, the other contour-hunting. The Golden Eagle also still-hunts, waiting on lakeside snags for passing watercraft to scare into flight waterfowl, which are sometimes snatched at the very bow of the boat.

In areas and at times of exceptional ground squirrel abundance, juveniles and subadults congregate in the fall, after local adult territory holders no longer enforce boundaries with much conviction.

REPRODUCTION: In the central Coast Ranges and likely elsewhere, at least at lower elevations, Golden Eagles begin their undulating courtship flights and dusk calling as early as November, gearing up for nest building and repair and February nesting. Through December and January, more pairs follow suit, the spectacular dives serving both to advertise the presence of a territory to other eagles and to stimulate the female to breed. Females, too, perform these dive displays, probably for territorial announcement. In the Golden Eagle, pinwheeling is likely territorial and performed with intruders, and therefore is not part of courtship.

Pairs appear monogamous, but if one member is lost, it may be replaced in a matter of days. Determined "homeless" females at times kill a territory holder and pair up with the victim's male.

In California, the great majority of Golden Eagles build their stick nests in oaks of various species, though other trees, such as western sycamores *(Platanus racemosa)* and gray and Monterey

Fig. 97. Juvenile Red-tailed Hawk killed and eaten by a Golden Eagle. The widely scattered feathers are typical of an eagle's meal.

pines *(Pinus sabiniana, P. radiata)* may be used as well. In the central Coast Ranges, optimal nest sites appear to be oaks partway down steep north- or east-facing slopes and in draws that shelter the nest from strong winds, but any other large tree in whatever location is acceptable when such ideal sites are already occupied or simply not available in prey-rich country. Almost always, however, the nest is below a ridgetop, allowing the eagle to glide down to the nest with heavy food items. A Golden Eagle builds on cliffs when such are available, as, for example, in San Diego County. On rare occasions, it simply places its nest on the ground.

Often, the nests are well concealed, and the adults may approach them flying under the canopy of the surrounding woodland; other nests are prominent because of their size and exposed location. A pair of eagles usually builds or repairs several nests

Fig. 98. A Golden Eagle nest in a western sycamore *(Platanus race-mosa)*. Golden Eagles through much of California are tree nesters.

each year before deciding on one. New nests are often surprisingly small for such a large bird, whereas old ones take on the proportions of stout towers from years of added layers of branches.

The quality of a nest depends on the skill of the builder; one subadult female, a first time breeder, appropriated a Red-tailed Hawk's nest and, instead of widening it, simply added to the top. The resulting platform was of such a small diameter that, during incubation, much of the female's body projected from the nest at odd angles; nevertheless, she managed to raise two young. Another female several times built nests of such poor quality that they eventually collapsed and fell out of the trees; on one such occasion, the half-grown young, having survived the fall, scrambled to the top of the slope above the nest tree and were raised there, on the ground, to fledging.

Typically two eggs are laid, and sometimes three or one, and very rarely four. Incubation lasts a little over six weeks and is chiefly the female's job. Upon hatching, the downy young grow very slowly, and an older chick may attack and kill its younger sibling. Many Golden Eagle pairs do not breed every year even though there may be no shortage of food.

The young remain in the nest for 10 to 12 weeks, and then depend on their parents for food for another two to three months or more.

DISTRIBUTION AND HABITAT: The Golden Eagle is holarctic in distribution. In North America, it breeds in Alaska and across much of Canada and is found year-round in the western United States, in north-central Mexico, and in northern Baja California. It winters in numbers in the southern plains states and sparingly in the southeastern states.

This species is found nearly the length and breadth of California in almost all habitats and can even sometimes be seen over cities. The presence and accessibility of prey is of the greatest importance, followed by availability of suitable nest sites. Large coniferous forests have few Golden Eagles; oak savanna interspersed with grassland has many. The inner Coast Ranges south of San Francisco offer this habitat combination in many places; southern Alameda County has the highest density of nesting Golden Eagles in the world. They occur sparingly in the deserts and the outer Coast Ranges the length of the state and are very rare breeders in the Central Valley. The Great Basin areas support a good population, and a few pairs are scattered across our deserts.

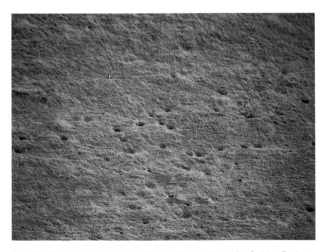

Fig. 99. Ground squirrel holes indicate abundant prey for Golden Eagles. Note grazed-down vegetation around burrows.

Golden Eagles can be surprisingly hard to find. Good places to look are Mines Road (southeastern Alameda County) and Hwy. 84 from Pigeon Pass (west of Livermore) to Sunol. Interstate 5 between Tracy and Los Banos is another productive area. Shasta Lake (Shasta County) and Lake San Antonio (San Luis Obispo County) provide good Golden Eagle viewing. Valleys that, in winter, host Ferruginous Hawks *(Buteo regalis)* are also good for Golden Eagles, as are roadside telephone poles in the Great Basin areas. Generally, this species prefers hilly country, and scanning the sky with binoculars in likely areas can sometimes reveal these birds soaring at great heights; they also commonly perch on high-tension towers in the right habitat.

SIMILAR SPECIES: In flight, a Golden Eagle has long, broad wings and a relatively long tail compared to a Bald Eagle *(Haliaeetus leucocephalus)* or Red-tailed Hawk. The narrow head is clearly visible from afar, unlike that of a Turkey Vulture, and sticks out farther than that of a buteo. However, it projects less than half the length of the tail, which distinguishes it from a Bald Eagle. A Bald Eagle does not fly with wings in a dihedral.

STATUS: A California Species of Special Concern, and a federal Bird of Conservation Concern outside California.

REMARKS: The Golden Eagle often is a creature of habit and likes to visit the same perches daily and loaf and sun itself, especially during the "off-season," when there are no young to raise. A Golden Eagle is generally retiring and does not allow close approach by humans, although there are notable exceptions, especially when there are newly hatched young in the nest.

Falcons

Falcons are generally thought of as hunters of the air; a large falcon cruising overhead is the epitome of raptorhood. As a group, they are generally the most streamlined of the raptors; the breast is often broad and muscular, the tail medium to long and tapering, and the wings sharply pointed. Most are relatively slow to accelerate but can sustain high-speed flight over long distances in pursuit of prey, unlike almost all other hawks. Surely one pinnacle of avian flight is the falcon's headlong power dive, the "stoop," that flings the hawk down from the sky in pursuit of prey.

A perched falcon looks nearly always trimmer than a buteo. Distinctive head patterns are frequently visible from afar. Most falcons have malar stripes, not to be confused with the often more diffuse malar patches of buteos. The eye is dark, very unlike the piercing yellow eye of an accipiter, and is surrounded by pale blue or yellow skin, which makes it more conspicuous. At close range, it lends the hawk an oddly mild, thoughtful air devoid of the fierceness seen in most other raptors. Falcons have generally long wings and short legs. Although all perch on high-tension towers, telephone poles, and fence posts, they often go unnoticed because of their habit of sitting on the ground as well, or on buildings, or flying at great heights.

Although the Crested Caracara *(Caracara cheriway)* is included in the falcon family, it is a scavenger and killer of very weak prey.

CRESTED CARACARA
(Pl. 4)

Caracara cheriway

RARE

IDENTIFICATION: The size is that of a slender Red-tailed Hawk. Its long neck, legs, and tail make this raptor look gangly. The tips of the folded wings extend to the tip of the tail.

The adult's head has a long black cap that ends in a spiky crest at the nape, white cheeks, and a bare face that can change from yellow to red. The large beak is pale bluish and yellowish. The neck is buffy white with black crossbars, the back and folded wings are dark brown. The throat is white, changing to buffy on the upper breast, which is narrowly barred black. The lower breast, belly, and leg feathers are black, and the undertail coverts are white. The spread wing shows the outer primaries white with faint gray bars and black tips. The white tail is faintly barred gray and has broad black terminal bands. The cere is yellow to red, the legs yellow.

The juvenile is like the adult but is dark brown where the adult is black. The cheeks and neck are buffy. Brown feathers of the upper back and breast have large buffy shaft streaks.

FLIGHT: The Crested Caracara flies with rather slow and steady wingbeats and soars on flat wings, the leading edges straight across from one wingtip to the other, creating an overall cross-shaped flight outline.

VOICE: In courtship and when excited, this species tosses back its head and gives a loud rattle call.

FEEDING: Although principally a scavenger, the Crested Caracara forages in a variety of ways. Where they are common, they are well-known habitués of roads, where they look for roadkill. But they also take injured birds and kill and eat turtles and dig up their eggs. Lizards and insects are caught by hunting on foot. In Mexico, one caracara was observed hanging upside down from an oriole nest and reaching in to extract the young.

REPRODUCTION: The Crested Caracara does not nest in California; elsewhere, it nests in palms, oaks, and other trees.

DISTRIBUTION AND HABITAT: A subtropical and tropical raptor of deserts, scrubland, and savannas, this species' normal range is well to the south and east of California, from some of the southern-most parts of the United States southward throughout most of Mexico and Central and South America. A very rare bird in California, it is an occasional visitor from Arizona and Mexico. Although most sightings are in southern California, it has been seen in other areas of the state as well, including the extreme north.

SIMILAR SPECIES: Crested Caracaras are virtually unmistakable.

REMARKS: Some sightings in California may be of escaped captives.

AMERICAN KESTREL
Falco sparverius
(Pl. 11; Figs. 30, 53, 100)

IDENTIFICATION: Smaller than a pigeon. The wings and tail are long. The wingtips of a perched bird do not reach the tip of the tail. This species often pumps its tail up and down several times while perched, and especially upon alighting. It frequently bobs its head up and down while looking at prey or potential enemies.

The adult's head has a blue gray cap normally with a rusty red crown; the sides of the head are white with two conspicuous vertical bars (one, the malar stripe) and, toward the nape, a black oval ringed with ocher. The sexes differ in body color and size (the female is a little larger). The underwing is finely barred and pale. The cere and feet are bright deep yellow.

The adult male has wings that are blue gray with some black spots, and the folded primaries are black with white tips. The back is rufous with transverse black bars on the lower half. The rufous tail has a broad black subterminal band and pale tip, but the outer rectrices are white and bear two black bars. The breast is pale rusty to cinnamon, with variable (but never heavy) black spots that become larger on the paler flanks. The belly is cream colored to white. In flight, the primaries have a row of white spots back of the trailing edge.

The adult female's back and wings are rusty brown with dark brown bars; the primaries are brown black. The rusty brown tail has many transverse bars, the subterminal one the widest. The breast and upper belly are beige to white with broad rusty streaks, contrasting with the unmarked lower belly and undertail coverts, especially in overhead flight, when these parts appear as a distinctly lighter oval.

A juvenile male looks like the adult but has a completely barred rufous back and scapulars and has fine black streaks on buff to white underparts. The juvenile female is like the adult female.

FLIGHT: In point-to-point flight, the long, pointed wings and long, narrow tail are obvious. The body appears slender, and the wingbeats usually are not snappy; rather, on the upbeat, it looks as if the wings are too heavy. Although the least dynamic of the falcons in its hunting behavior, this little falcon can at times fly very fast, especially when pursuing birds, and then can easily be mistaken for another species (as in poor light when color is not easily seen). Very often, it hovers with fluttering wings and tail partly or wholly spread, or kites into a strong wind. In a soar, its shape can appear unfalconlike, the wings can look much less pointed, and the tail may be completely fanned out so that it looks very broad. The belly appears as a pale oval.

VOICE: The American Kestrel is fairly vocal and produces at least three distinct calls, the most frequent being the "killy" call used when mobbing another predator and during other excitement such as courtship. Whining is associated with begging for food by both adults and young, and chittering is heard from both sexes during nesting activity.

FEEDING: The two main foraging methods of this falcon are still-hunting and hovering or kiting over cover likely to hold prey. In either case, the raptor drops on or angles down to the quarry, usually descending headfirst, but sometimes also "parachuting" down, feet first, if the prey is slow moving. In addition, the American Kestrel soar-hunts in pursuit of large insects, especially dragonflies, which it grabs with its feet and consumes in midair, having first removed the wings, which drift down like glittering flakes. For desert kestrels, lizards and small snakes form a substantial part of the menu.

In winter, this falcon, especially the male, may turn to small birds when insects become scarce, particularly in urban settings. Bird hunts usually are launched from a distance, the hawk steadily accelerating as it shoots, a few feet off the ground, toward a flock of birds feeding on the ground, such as House Sparrows *(Passer domesticus)* or European Starlings *(Sturnus vulgaris)*, nabbing one as they scatter. Some specialists can even capture birds feeding in treetops, such as pine siskins *(Carduelis pinus)*. The male American Kestrel is also quick to take advantage of the half-grown young of various swallows waiting to be fed in the entrances of their nests, and of awkward swallow fledglings on their maiden flights, these morsels being conveniently abundant when the kestrel has young of its own. Unlike a Merlin *(Falco columbarius)*, it is, however, not given to long chases after a bird, choosing to catch it on the ground or perch or as it rises.

By preference, the American Kestrel would likely be happy with a steady diet of grasshoppers and crickets, though mice and voles are taken year-round where available and constitute the chief food for the young, insects holding inadequate amounts of water for the growing chicks.

REPRODUCTION: American Kestrels in California begin to show reproductive behavior as early as February. The female solicits the male by flying about or perching with quivering wings that arch downward, and the male delivers food to the female and to the

Fig. 100. Male American Kestrel perched on snag above its nest hole.

nest with a similar flight. This display appears identical to the food begging flight of fledged young. The male also performs pendulum dives over the perched female, quivering his wings at the highest point of each arc and calling. The male's skydance consists of undulating flights, a series of spectacular power dives and pull-ups over a distance of perhaps 100 m (300 ft).

No nest is built, the birds making do with whatever substrate is found in a suitable cavity. Former flicker holes in snags are most commonly used, but a wide variety of other chambers are acceptable, such as holes in cliffs, dirt banks, and buildings, old magpie nests (which are roofed), the dead fronds below the crown of a palm, crannies amidst bridge girders, and nest boxes, often those intended for Wood Ducks *(Aix sponsa)* and Western Screech Owls *(Otus kennicottii).* The hawk scrambles into the nest entrance surreptitiously and at speed, so fast that the observer often is uncertain what has occurred. In years of great prey abundance and in areas supplying sufficient nest sites, American Kestrels may nest in small, loose colonies.

The American Kestrel lays three to five eggs, with four being the most common, and incubation takes a little over four weeks. The young fledge after another month and become independent of their parents after a few weeks. They then may join in social hunting groups with other broods while the parents may breed again and raise a second set of young.

DISTRIBUTION AND HABITAT: The American Kestrel is common and widespread in North America, although northern populations withdraw during the winter. Its range extends through central and western Mexico, with isolated populations in Central America, and the species is widespread in South America.

In California, one of the most widespread and numerous of all our raptors, this charming little falcon is commonly and easily spotted along roadsides, often perched on telephone lines, poles, snags, and light standards, or hovering or kiting over center dividers and roadside berms. From coastal bluffs to the meadows of mountain forests to the washes of California deserts, this kestrel takes advantage of available prey. Migrants augment the California population during the winter months, and California birds from areas with an inhospitable winter climate move into the interior valleys and deserts, increasing numbers there.

The American Kestrel avoids forest interiors. It is especially fond of savanna and oak woodland but can be expected in every other habitat, including city centers, except over large bodies of open water.

SIMILAR SPECIES: A Merlin glides much less than an American Kestrel and normally shows no deep rufous; it also has a much snappier wingbeat and does not hover.

REMARKS: This kestrel is a demonstrative, sometimes noisy hawk that frequently allows close approach, especially where its chosen habitat happens to be an urban environment. American Kestrels eat great numbers of insect pests and mice but are themselves occasionally eaten by larger raptors or hit by cars when pursuing prey that is crossing a road.

MERLIN
Falco columbarius

(Pl. 11; Figs. 14, 48, 101)

IDENTIFICATION: In size (but not in shape), a Merlin is like a small, slim pigeon. It is small headed for a falcon, slender, and long tailed. The wingtips of a perched bird reach two-thirds of the way to the tip of the prominently barred tail. It may slowly pump its tail once upon alighting, and sometimes sits with the tail fully fanned, especially when warming up or when drying out on damp mornings (see pl. 11). Often, it is fidgety while perched. The sexes differ in color (in adults) and size, the male being smaller.

Three well-marked races appear in California, differing in color saturation, and some individuals are intermediate in color. The Taiga race *(F. columbarius columbarius)* is the most common form seen in California. The Prairie race *(F. c. richardsonii)* is the palest. The Black race *(F. c. suckleyi)* is the darkest (may appear all black).

The adult Taiga race male has a dark blue gray crown, buffy and streaked cheeks, and a conspicuous pale superciliary. The malar stripe is narrow and broken. The back and folded wings are dark blue gray, the feathers with thin black shaft streaks. The throat is conspicuously white, and the underparts are white to buffy with thick, dark brown streaks, the leg feathers and undertail coverts bright buffy to cinnamon colored. The underwings have a striking checkerboard pattern, especially on the flight feathers. The tail is black and has three narrow gray to white bands and a white terminal band.

The adult Taiga race female has crown and cheeks variably brown, with pale superciliaries. The malar stripe is broken and narrow. The back and folded wings are dark brown. The whitish throat is conspicuously set off from the dense dark brown streaks on the white underparts. The leg feathers and undertail coverts are pale buff and finely streaked. The tail is dark brown with three narrow whitish bands and a white terminal band.

The adult Prairie race male is like the Taiga male but much paler; the light-colored nape makes the bird look collared. The streaking on the underparts is rufous brown. The black tail often has four whitish crossbands, wider than those of the Taiga male, plus a white terminal band.

The adult Prairie race female is like a faded Taiga female. It has a tan to reddish brown crown, back, and folded wings. The streaking on the white underparts is brown to light rufous. The brown tail has four or five pale bands, wider than in the Taiga female, and a white terminal band.

The adult Black race male has a head appearing virtually all black, without a pale superciliary or malar stripe. The back and folded wings are slate black. The underparts are almost entirely brown black, the streaks so wide they almost obscure the rufous base color. The rufous leg feathers are thickly streaked, and the undertail coverts barred. The tail is black, the pale bands often entirely absent or reduced.

The adult Black race female is like the male, but dark black brown, without a slaty cast, and the streaked leg feathers are buffy instead of rufous. The streaking on the underparts may nearly obliterate the pale background.

In all races, juveniles of both sexes are like the adult female. The Prairie race juvenile male may have underparts streaked cinnamon.

FLIGHT: In flight, a Merlin appears small and fast although females often look larger than they really are. The wingbeats are snappy, and the hawk always seems to be in a hurry. The pointed wings are flat in a soar, with the tail partly or fully spread. The underwings are dark, with a striking checkerboard pattern.

VOICE: On their breeding grounds, Merlins can be very noisy, especially when they have young; one of their vocalizations, a strident staccato "yick-kyic-kyic" is given in nest defense and also, on occasion, in attacks on other raptors while on migration (as can be heard in the 1987 movie *The Untouchables,* in the bridge scene

Fig. 101. Although usually sparrow hunters, Merlins sometimes catch surprisingly large quarry, such as this Mourning Dove.

at the Canadian border). Although Merlins are probably the noisiest of all the falcons, by and large, when they are in California and far from their breeding grounds, they are generally silent.

FEEDING: Intense and alert, a wintering Merlin in this state seldom sits still for very long unless it has recently eaten, and even then, its little head keeps turning as it keeps an eye out for predators and watches other birds fly by. In early morning, this falcon often seeks out the top of a snag or tree as a foraging post for still-hunting; later in the day, having eaten, it hides in the interior of leafy or bare deciduous trees. The Merlin is also frequently seen prospecting, skimming along at speed over salt marshes, mudflats, grasslands, and sometimes backyards in hopes of surprising prey. Along bay shores, these falcons can be seen making spirited stoops at wheeling flocks of shore birds, especially Dunlins *(Calidris alpina)*. One female accompanied three different Northern Harriers *(Circus cyaneus)* in succession as they quartered over grassland, flying oval paths about 50 m (160 ft) above each for a total of 42 minutes, and stooping nearly vertically at Savanna Sparrows *(Passerculus sandwichensis)* flushing ahead of the hawks; she failed to catch one and finally went to perch, but it was clear from the falcon's persistence that this was often a successful ploy, and indeed, this behavior has been reported several times elsewhere.

A Merlin may pursue quarry into cover and can run quite well and is reminiscent of an accipiter. Birds are the principal food of this falcon, and a great variety is taken, particularly species that inhabit open areas for example the smaller sandpipers, plovers, pipits, larks, various sparrows, and swifts and swallows. Mourning Doves (*Zenaida macroura*) and even Rock Doves (city pigeons, *Columba livia*) are occasionally caught, the latter usually by individuals that specialize in this quarry so monstrously large for a tiny falcon. It has been stated that the Merlin kills such big prey by strangulation, but that is debatable; although it is true that it habitually wraps a foot around the neck of any bird, it always follows up with bites to the neck in typical falcon fashion in order to sever the spinal cord. Mammals such as mice and bats are also occasionally caught, as well as large insects, including numerous dragonflies, which are commonly eaten on the wing.

REPRODUCTION: As might be expected, courtship activity tends to be exuberant, with a variety of aerial displays that include flying with deep, strong wingbeats and lateral rolls by the male (power-flying) and power-diving. Unfortunately, the Merlin does not nest in California, and the Merlin watcher has to travel as far as Montana and Wyoming to see any of these carryings-on. Like other falcons, the Merlin does not build a nest, but in the western United States most commonly uses the vacant nests of large birds, principally magpies and crows.

Incubation of the four to six eggs takes four and a half weeks, and the young leave the nest at about four weeks of age.

DISTRIBUTION AND HABITAT: A circumpolar breeder of the northern Holarctic, in North America the Merlin breeds across most of Canada and Alaska and southward into the northern Rocky Mountain states. It winters west of the Rocky Mountains, along the east coast, throughout Mexico and Central America, and in the northernmost part of South America.

Merlins occur in numbers throughout California during winter, but because of their small size and fairly secretive habits, they are easily overlooked. The majority arrive fairly late in October and November. The Prairie race tends to be seen more often in the Central Valley and the Black race on the coast, but these less common forms can show up anywhere.

Merlins are clearly drawn to areas holding a good supply of birds that are sufficiently exposed to make capture possible. Bay marshes and grasslands seem popular, as well as valley farmlands, dairies, savannas, and edges of deserts; some stay in cities during their winter visit and with a little practice can be spotted atop trees such as coast redwoods *(Sequoia sempervirens),* deodar cedars *(Cedrus deodara),* and eucalypts (*Eucalyptus* spp.). But they have also been seen at elevations ranging from 1,200 to 2,400 m (4,000 to 8,000 ft) in winter, pursuing birds in the treetops (J. Schmitt, pers. comm. 2004).

To find a Merlin, snags and trees with horizontal branches at their very tops must be scrutinized, especially very early in the morning, and the interior of deciduous trees later in the day, after the falcon has likely fed; obviously, this is much easier after the leaf fall, though Merlins arrive well before then. Merlins are seen in the Klamath Basin and other generally good raptor areas in the northeast (such as the Ash Creek Wildlife Area in Modoc County), in coastal areas with lots of shorebirds, and in the Sacramento, Owens, Antelope, and southern Santa Clara (Santa Clara County) valleys. They are also commonly found in the Ramona Grasslands (San Diego County); all three races reach that southernmost county of the state.

SIMILAR SPECIES: The similarly sized American Kestrel *(Falco sparverius)* often sits on thin twigs or wires (very rare perches for a Merlin), and it frequently and deeply pumps its tail numerous times upon alighting. Perched, if the hawk is fat around the middle and has a really large head, it is almost certainly a kestrel. A kestrel, especially in cool weather, is frequently fluffy, with a hunched back; it has two conspicuous vertical black bars bordering white cheeks. In flight, the kestrel is floppier, as if its wings were heavy.

STATUS: A California Species of Special Concern.

REMARKS: Unlike the small accipiters, which are masters of the short dash, the Merlin generally likes a longer in-run for its attacks and is therefore more often found on open ground, for instance where windbreaks of tall trees border agricultural expanses or playing fields; parks with open areas and large trees are also popular. However, it is quite prepared to hunt over backyards and can therefore be observed at times in street trees. Some trees are such irresistible magnets that they are sought out by Merlins year after year during the winter months, sometimes by the same hawk, at other times by new ones. In salt marshes, a Merlin may sit on fence posts for want of a better perch. This species also perches at times on weed stalks inches from the ground.

GYRFALCON
Falco rusticolus
(Pls. 11, 12; Fig. 21)   RARE

IDENTIFICATION· Our largest falcon, Redtail sized, appearing broad chested and rather nondescript. It frequently perches with its legs hidden by feathers (the toes sometimes show, however). The eye appears small in the puffy head, and the malar stripe, which may be entirely absent, is usually faint or blurred or both because of surrounding fine streaks. The wingtips of a perched bird are well short of the tip of the tail.

The three morphs of the Gyrfalcon (the white, gray, and dark) each have distinct adult and juvenal feathering; however, only the gray form has been confirmed in California. In all, the sexes are similar in pattern and color, but the males are smaller, though often not as dramatically so as in other falcons. The adult of all forms has yellow cere and legs.

The adult gray morph has a gray head with a gray, poorly defined malar stripe, streaked cheeks, and a pale forehead. The back and folded wings are gray with numerous paler gray crossbars, the projecting primaries gray tan with indistinct barring. Underparts are white with many small blackish spots, larger on the sides and becoming bars on the flanks. Leg feathers and undertail coverts are barred gray. The tail has numerous gray black bars equal in width to the paler bars between. The underwing coverts are white with dark streaks, the barred flight feathers darker and gray. In flight, the underparts appear uniformly pale.

The juvenile gray morph has a head streaked brown on white with a pale superciliary line, an indistinct dark brown malar stripe, and a pale, streaked cheek. The back and folded wings are gray brown, the feathers with pale, whitish margins. Underparts are off-white with broad, dark brown streaks. Leg feathers are streaked. In flight, heavily marked underwing coverts contrast with much lighter, faintly barred flight feathers. The cere and legs are bluish to yellowish.

FLIGHT: The Gyrfalcon has several speeds, the most surprising (for a large falcon) being a low gear for cruising very slowly, appearing in fact slower than a Redtail *(Buteo jamaicensis),* with leisurely shallow wingbeats followed by stately gliding. It has at least two higher speeds, the fastest for active pursuit flights, employing fast, shallow wingbeats. A Gyr soars with wings flat, and sometimes it may hover. In flight, it looks heavy bodied, with a long, wide tail and tapering wings that appear very wide at the base. The body is ample across the middle and tapers both toward the head and the tail.

VOICE: This falcon is generally quiet, vocalizing mainly during the breeding season and is therefore not likely to be heard in California. The calls are very similar to the calls of other large falcons.

FEEDING: The Gyrfalcon hunts chiefly from a perch and by employing speculative flights, skimming along at low altitude hoping to surprise quarry. It also attacks in a stoop but does so much more rarely than the Peregrine Falcon *(Falco peregrinus),* relying instead on its great speed to outfly birds in a tail chase. Overtaken prey is commonly knocked down first, and then seized on the ground. Also unlike a Peregrine or Prairie Falcon *(Falco mexicanus),* which need to circle to gain height, a Gyrfalcon can power up in a straight line and at a steep angle; it is an extremely strong flier.

The Gyr takes many mammals, ranging in size from mice to Arctic Hares *(Lepus arcticus);* ground squirrels are a common prey. Birds, another major food, range in size from finches and sparrows to geese.

REPRODUCTION: Courtship displays are very similar to that of the other large falcons. Cliff ledges are probably preferred nest sites, but where unavailable, the falcon readily uses old stick nests built by other birds.

Typically four eggs are laid, hatching after five weeks of incubation. The young fledge after about seven weeks.

DISTRIBUTION AND HABITAT: The Gyrfalcon is a circumpolar nester of Arctic regions, in winter in North America moving southward all across Canada to the United States border and variably beyond in small numbers.

It is an extremely rare winter visitor in California and has turned up in only a handful of places. It has probably never been seen south of the San Francisco Bay Area, and all sightings appear to have been inland, with the exception of bays to which the falcon is attracted by abundant prey. It has also been observed on agricultural land and at large freshwater marshes.

More visit in some years than in others. The best chance of seeing one in California is in the Klamath Basin near Tule Lake. The Gyr is fond of telephone poles.

SIMILAR SPECIES: A Peregrine Falcon has a darker head, a more defined face pattern, and longer wings. A Prairie Falcon is browner, looks longer winged, and has dark armpits. An adult Northern Goshawk *(Accipiter gentilis)* has a uniquely marked head, and in flight its wings usually are not as pointed as a Gyrfalcon's.

REMARKS: Gyrfalcons are much favored by some falconers because of their intelligence, playful nature, and willingness to bond, doglike, with the owner. It is therefore possible that a Gyrfalcon seen in California is in fact a lost falconry bird, although Gyrs are exceptionally valuable and not usually lost for long. The gray form is the most commonly trained, compared to the much scarcer white and dark morphs.

The adult white morph is entirely white with dark wingtips, or has few scattered brown to dark brown spots or bars on the back and folded wing; the juvenile is white with brown or blackish bars and streaks on upperparts. The adult dark morph has a dark blue gray head, back, and folded wings and is more heavily marked on the underparts; the juvenile is entirely blackish brown except for a scattering of whitish spots and streaks on the underparts. Its underwing coverts are very heavily marked blackish brown and contrast with the much paler tan flight feathers; the tail has numerous broad black brown bands. Dark individuals may have pigmentation so saturated that they appear black from a distance.

A great variety of intermediates exists between these forms. Gray white intermediates are sometimes called "silver Gyrs"; they have mostly white heads and breasts but gray backs. To compound the problem, captive-bred hybrids are common.

Contrary to many dictionaries, the common name of this species is usually pronounced "jeer'-fal-con."

PEREGRINE FALCON   *Falco peregrinus*
(Pls. 11, 12; Figs. 11, 23, 24, 49)

IDENTIFICATION: Larger than a crow but smaller than a Red-tailed Hawk. Its head often appears rather small and rounded and has a broad, always conspicuous malar stripe or is all black. The wingtips of a perched bird reach the tip of the tail, or nearly so. The legs are short. The feet are large, grayish to deep yellow, and the cere is yellow.

Three races of Peregrine occur in North America, all of which appear in California. They differ mainly in the width and extent of the malar stripe and the degree of marking on the undersides. The lightest form is the Tundra *(F. peregrinus tundrius),* followed by the Anatum *(F. p. anatum),* which is the most commonly seen race in California; the Peale's Peregrine *(F. p. pealei)* is darkest. The sexes generally look alike, but the male is smaller and less heavily marked ventrally.

In all adults, the back and folded wings range from bright blue gray to blackish gray to a dull ashy gray, usually showing a gradation from darkest at the neck and mantle to lightest at the upper-tail coverts. The breast looks like a white to buffy bib, at times with thin streaks or small spots (heavily spotted in the female Peale's). Flanks, belly, leg feathers, and undertail coverts are conspicuously barred. The black tail has numerous gray bars and a white terminal band. Other differences in the races are as follows.

The adult Anatum's head is blackish, with a broad malar stripe that may invade or even cover the buffy cheek, making the entire head black. The belly is often buffy to nearly salmon colored.

The adult Tundra has a blue gray to blackish head, with a narrow to medium wide black malar stripe that contrasts sharply with the clean, white cheek. Birds of this race are usually more slightly built than the other forms, a trait that is not readily apparent unless the falcon is very close at hand.

The adult Peale's has a blackish head with an ashy gray cast and broad malar stripe. The female has a dirty-looking cheek that is streaked with gray black, a breast that is heavily spotted, and thicker barring on the flanks, belly, leg feathers, and undertail coverts.

A juvenile Peregrine Falcon's head usually has white or buffy superciliaries that connect in back of the head and isolate the brown crown (as in the Tundra and some Anatums), or the head is dark brown. Underparts have conspicuous streaking, broadest in the Peale's. The folded brown to blackish tail may have numerous brown crossbars (but none in the Peale's), with a whitish terminal band. In flight, a juvenile Peregrine looks very long tailed.

The juvenile Anatum has a wide malar stripe. The cheeks are buffy, often streaked. The back and folded wings are dark brown with variable buff or with rufous edging on the feathers. Underparts are buffy to rufous.

The juvenile Tundra has a head with a whitish to buffy forehead and conspicuous superciliaries. The malar stripe is narrow, and the cheek white or pale buff. The back and wings are brown with variable buff or rufous edging. The pale buff underparts are narrowly streaked.

The juvenile Peale's lacks all buffy and rufous tones and looks black from a distance. The cheek is heavily streaked. The back and wings are blackish without edging. The folded tail lacks bars. Underparts have a whitish base color with conspicuous broad streaks, and streaks cover the throat to the beak.

Intermediates between these forms occur.

FLIGHT: A Peregrine flies directly, with no hovering. Deep, rowing beats alternate with gliding. It soars with wings flat, the tail fanned. The underwings are heavily barred and appear uniformly dark. The wings are usually very pointed looking unless the bird is soaring; this falcon appears long winged and long tailed in the air. Juveniles, especially, look very long tailed in flight.

VOICE: The Peregrine is generally silent but highly vocal during the breeding season, when "ee-chup" calls are used for greeting between the partners and throaty "kak-kak-kak" notes announce the presence of a predator. A variety of other calls is associated with begging, copulation, and other breeding activities. Many vocalizations are very loud and carry great distances, and they can be heard over surf crashing on rocks.

FEEDING: More than any other raptor, the Peregrine Falcon is specialized for catching flying prey, usually birds. Although other hawks drive their quarry into cover where it has limited mobility, or snatch earthbound animals, a Peregrine much prefers its victim be on the wing. If the prey is small, it may even be plucked and eaten in flight. Thus, when a Peregrine still-hunts, it is scanning the sky from some perch for suitable quarry that may be a mile or more away. In addition, this falcon also uses speculative flights, taking advantage of the contours of the land in hopes of startling a bird into the air, or it attacks from a high flight or a soar.

Such hunts initiated from a height take the form of the celebrated stoop, the meteorlike plunge downward during which the falcon's speed may exceed 380 km (240 mi) per hour (Franklin 1999). Pulling out of such a dive, the falcon is often propelled upward 30 m (100 ft) or more and is repositioned for a second try. Some prey birds adroitly dodge the Peregrine's first pass, and then follows a series of stoops, usually diminishing in height. If the struck quarry has fallen to the ground, the falcon settles on it and, if still alive, dispatches it with bites to the neck that usually sever the spinal cord. Although some birds are caught in straight tail chases, these level flights are usually initiated from a height, and, normally, a Peregrine endeavors to get above the quarry; having missed its stoop, the falcon can also shoot upward, carried by its own momentum, and seize the prey from below.

Although Peregrines catch some insects and sometimes bats, midsized birds are the chief prey; pigeons and doves are commonly taken, but essentially, this falcon catches what is locally available. One California eyrie had an abundance of legs of Black-necked Stilts *(Himantopus mexicanus)* (G. Monk, pers. comm. 2002), another had numerous remains of European Starlings *(Sturnus vulgaris)* and Western Meadowlarks *(Sturnella neglecta)*. The size range of recorded prey birds is great, from wrens to cormorants and geese.

REPRODUCTION: As befits so aerial a hunter, courtship flight displays are spectacular. The male flings himself about the sky in steep undulations; both partners circle high and engage in various interactions, including lateral rolls as they dive at each other, even the touching of beaks.

Eyrie sites are defended with vigor against other Peregrines or other large falcons; in one instance, a female Peregrine struck dead an intruding female Prairie Falcon *(Falco mexicanus)* in a devastating stoop (Walton 1978). Such resolute defense is necessary, because Peregrines nearly always require a cliff, a relatively scarce commodity in California. However, if prey is plentiful, this falcon sometimes utilizes large stick nests of a Red-tailed Hawk *(Buteo jamaicensis)*, Common Raven *(Corvus corax)*, or other birds built in sandy coastal bluffs (White et al. 2002), and one old record exists of a nest in a barrel in a Palo Alto (San Mateo County) salt marsh. Although the largest population in California is in the Coast Ranges foothills, with nesting cliffs 15 to 120 m (45 to 360 ft) high, on the Channel Islands this species sometimes nests on the ground in gull colonies (Walton 2003). Still other eyrie sites include human-made structures (usually provided with an artificial nest box), such as bridges and high-rises (Bell et al. 1996); unfortunately, fledging success is much lower from these nest sites, especially bridges. In the northwest of the state, Peregrines sometimes use high hollows in large trees, as well as sea-stacks (nearshore rock pinnacles that project from the sea). Very often, there is water near the nest cliff—the ocean or a stream below or perhaps a marsh a few hundred meters away. Ledges with overhangs and potholes are preferred, typically though not always near the top of the cliff (see fig. 73); the Peregrine nest ledge on El Capitán in Yosemite National Park is at

1,100 m (3,600 ft.) far below the summit. Often, there is vegetation on or near the ledge.

No nest is built, and the three or four eggs are laid in a simple depression or scrape fashioned by both partners. The eggs hatch after four to nearly five weeks, and the young leave the nest when about six weeks old.

DISTRIBUTION AND HABITAT: With the exception of Antarctica, the Peregrine Falcon is cosmopolitan. In North America, this falcon breeds across the Arctic, in the Rocky Mountains, and along the northeastern seaboard. It is a year-round resident along the mid-Atlantic coast, along nearly the entire Pacific coast, and through Baja California and much of mainland Mexico. Scattered breeding pairs are found throughout the United States. Wintering areas include much of California and the coastal strips of the southeastern states.

In general, the Peregrine has a dynamic distribution and shows up sometimes in surprising places. It can be found throughout California but very rarely in the deserts; it does appear occasionally at the Salton Sea and the lower Colorado River Valley and even nests in the deserts of Baja California. It is most often seen along the coast but also occurs in the high mountains, the Central Valley, and in the foothills and Coast Ranges. The breeding habitat requirements of a large cliff or similar structure, usually near or above water, and a good supply of avian prey limit nesting to wherever these conditions are met, most commonly along the coast, but also locally in the Coast Ranges, the foothills, and the Sierra Nevada.

During the winter months, there is a substantial influx of Peregrines from farther north, and a greater variety of habitats is utilized. Because of the presence of great numbers of potential prey birds, wintering Peregrines seek out bay shores, salt marshes, and estuaries, freshwater marshes of the Central Valley, and dairy country as well as cities.

Despite their rapid increase in numbers since the 1980s, Peregrines are elusive. They seem to prefer exposed perches; at least, that is where they are usually seen when they are seen at all. The most likely places to see one include the wetlands surrounding San Francisco Bay (scrutinize the high-tension towers in fall, winter, and to some extent spring), especially on the eastern end of the San Mateo Bridge (Alameda County) and at Dixon Landing (Santa Clara County), the mouths of the northwestern rivers, the coastal bluffs near San Francisco and Santa Cruz and the Big

Sur coast, and during the nesting season at Morro Rock (San Luis Obispo County). They winter and breed in some California cities but are not easy to see there.

Although the Anatum is the resident race in California, both Tundra and Peale's forms appear in fall and winter. Northern populations of Anatums are migratory, but those that breed in California usually do not stray far from their breeding territories.

SIMILAR SPECIES: A Prairie Falcon has dark armpits in flight; it tends to sit at a slight diagonal, has a blocky head with a narrow malar stripe, and shorter wings. A Prairie also has a distinctive light area behind the eye. A Gyrfalcon *(Falco rusticolus)* is much bigger and heftier, lacks the distinctive cap of the Peregrine, and has shorter and wider wings and a longer tail.

STATUS: Only the Anatum race is listed as endangered in California. Federally, the Peregrine is now a Bird of Conservation Concern.

REMARKS: The restoration of the Peregrine Falcon to the California landscape after its catastrophic decline has been nothing short of spectacular; ironically, its increased presence threatens the endangered California Least Tern *(Sterna antillarum browni)*, which happens to often nest in areas patrolled by this falcon.

Fig. 102. Hybrids produced by artificial insemination (often Peregrines crossed with other species) can look very exotic and confusing. *Left:* Gyrfalcon × Peregrine, black type (a very popular cross). *Right:* Peregrine × American Kestrel (an experimental cross). The first-listed species denotes the semen donor. (Not to scale.)

PRAIRIE FALCON   *Falco mexicanus*

(Pls. 11, 12; Figs. 7, 44, 103)

IDENTIFICATION: Larger than a crow but smaller than a Red-tailed Hawk and slightly smaller than a Peregrine *(Falco peregrinus).* A Prairie Falcon is stocky and short legged, with a large blocky head that has a conspicuous white or pale superciliary line that continues around the back of the head, isolating the brown crown, and a narrow but distinct malar stripe. The tips of the folded wings do not reach the tailtip.

The sexes are nearly alike in color and pattern, but the male is smaller than the female. A perched bird (especially the female) often folds its tail to one feather width so it appears very narrow when seen head-on or from the back.

The adult's head has two-toned cheeks, the upper half brown and the lower part white or pale buff. A conspicuous, small, pale wedge is behind the large eye. Upperparts are tan or gray brown, with narrow pale barring. Underparts are white with very small, short, drop-shaped spots that get larger on the flanks, much larger in the female. Leg feathers have spots and posterior bars. The gray brown tail has many pale bars. When seen from above in flight, the tail is distinctly lighter than the remainder of the upper side, with a noticeable pale rufous tone. The cere and feet are orange yellow.

The juvenile is like the adult, but darker brown on the head, back, and folded wings. The underparts are buffy, fading to white by winter's end, with conspicuous, dark brown streaks that widen on the flanks. The flight feathers of the wings and tail are more conspicuously narrowly barred. The cere and legs are blue gray, turning yellow in spring.

FLIGHT: A Prairie Falcon flies with shallow, fast wingbeats, snappy in the male and powerfully rowing in the female. It glides with wings bowed downward and soars with wings held flat and its tail partly fanned. It hovers rarely and briefly. Females, especially juveniles, flare out the feathers of the flanks, legs, and belly when landing or taking off, so that they appear to be wearing a full skirt.

In flight, the pointed wings are narrowly barred, the axillaries conspicuously black brown, this color continuing outward on the underwing coverts bordering the flight feathers. The tail is long, and in the adult, it is often paler and pinker than the rest of

Fig. 103. Prairie Falcon, showing its full "skirt."

the upperparts. The juvenile in flight is similar to the adult, except the back is browner and the tail slightly longer.

VOICE: A Prairie Falcon rarely vocalizes outside of the breeding season. Among the calls given near the eyrie are a series of harsh "kak-kak-kak" notes of high alarm and also wails during copulation.

FEEDING: The most commonly seen foraging methods are still-hunting (with long, ground-level dashes at prey), speculative low flights over promising cover, and long, shallow stoops from a height or from high-tension towers. A Prairie Falcon's bag of tricks, however, is impressive. An adult male chased a Brewer's Blackbird *(Euphagus cyanocephalus)* into a riparian woodland of

Fig. 104. Pinnacles National Monument. The rocks provide many eyrie sites for Prairie Falcons and perhaps Peregrines.

willows; having failed to catch it, he continued through the tree-tops for about 100 m (300 ft) in an apparent effort to flush something, then emerged and circled up in the air. After a few minutes, the falcon made a long, very shallow dive at a flying group of six blackbirds, which wheeled out of the way and then shadowed him for a short distance, keeping safely slightly above and behind; next, he dropped down to a weed-choked fencerow and followed it for about 150 m (500 ft), popping over the top periodically, again in an apparent effort to flush something. The Prairie then repeatedly buzzed four White-tailed Kites *(Elanus leucurus)* perched near one another on fence posts, seemingly in hopes of relieving them of any quarry they might have. Finally, he circled up again to about 75 m (250 ft) and then dropped into a moderately steep dive until he was a meter or so above a cut-over alfalfa field; without a wingbeat, he shot along at that level for another 50 m (160 ft) or so, and then deftly snatched a small bird off the top of a bare stalk, pulled up abruptly, and landed with his quarry. The victim, probably a Savanna Sparrow *(Passerculus sandwichensis)* likely never saw the falcon coming.

Besides open-country sparrows, the Prairie Falcon is very fond of Horned Larks *(Eremophila alpestris),* meadowlarks, and various ground squirrel species. It also takes starlings and other similar-sized birds, doves, pigeons, pheasants, and various duck species. In some places, lizards form part of the diet, as well as insects. It is an exceptionally hardy, rough-and-tumble raptor and is quite prepared to dive into cover after quarry and engage in a ground battle. Female Prairies are known to catch full-grown California Jackrabbits *(Lepus californicus)*, prey that may weigh three or four times as much as they do and that they basically bash to death by repeated blows with their feet, delivered in short, slashing dives.

REPRODUCTION: During courtship, the male often powers up to great heights above its mate and performs a series of spectacular stoops, and both male and female fly back and forth in front of the nest site, vocalizing.

The Prairie Falcon is a cliff nester; like other falcons, it builds no nest but instead uses existing facilities, such as potholes, ledges, and old nests of Red-tailed Hawks *(Buteo jamaicensis)* or Common Ravens *(Corvus corax)*, nearly always in the upper third of the cliff, and with an overhang. In contrast to a Peregrine Falcon, a Prairie at times uses rather insignificant rocks stuck away in small canyons (especially in chaparral country), some no higher than a dozen meters or so, and even dirt banks. Very large cliffs are sometimes shared with nesting Common Ravens, Red-tailed Hawks, and even Golden Eagles *(Aquila chrysaetos)*.

Most commonly five eggs are laid and incubated for about four and a half weeks, and the young leave the eyrie at approximately five and a half weeks of age.

DISTRIBUTION AND HABITAT: This strictly New World falcon breeds west of the Great Plains from southern Canada into Mexico. In most of that area it is found year-round; wintering grounds include the Great Plains, the Pacific coast, and northwestern Mexico.

Although the Prairie Falcon occurs nearly throughout the state, it clearly prefers the drier parts for nesting; only one instance of a coastal eyrie is known, from southern California. The species is not found in coniferous forests, but it has been found nesting above 3,600 m (10,000 ft) (B. Walton, pers. comm. 2002). The densest breeding populations appear to be in the inner Coast Ranges, parts of the eastern foothills of the Central Valley, and the arid northern and northeastern parts of the state. Smaller numbers breed in California's deserts and arid lands east of the Sierra Nevada, such as in the Owens Valley. Nesting habitat includes grassland, oak woodland (with nearby grassland, chaparral, or both), sagebrush, and desert scrub, always with the required cliff or high bank for a nest site. Pinnacles National Monument (San Benito County) has on average 12 nesting pairs.

Wintering Prairie Falcons additionally utilize agricultural areas, coastal plains and bay salt marshes, and grasslands; unlike Peregrines, they normally do not take advantage of urban habitats. Some may show up in late summer above timberline.

Prairies perch atop oaks in savanna, on high-tension towers in grasslands, on telephone poles, and sometimes on fence posts, flushing easily when approached. They often sit on the ground, where they may go unnoticed.

Although there are substantial numbers in the drier parts of the Coast Ranges south of San Francisco, they manage to make themselves scarce. Good places to look are valleys such as the Livermore Valley (eastern Alameda County), the foothills of the inner Coast Ranges from Tracy (San Joaquin County) to Coalinga (Fresno County), the Klamath Basin and Modoc Plateau in the north, and Lake Henshaw (San Diego County). You have a good chance of spotting a Prairie Falcon at Pinnacles National Monument.

SIMILAR SPECIES: A Peregrine often appears more slender, longer tailed, and longer winged, has a more conspicuous facial mask, usually lacks a pronounced superciliary, and lacks dark armpits. A juvenile Peregrine has a darker back. A juvenile Gyrfalcon *(Falco rusticolus)* is much bigger and usually has coarser and paler streaks on its underparts. A Red-tailed Hawk sits straighter, usually has a thicker waist seen from afar, and has an even bigger head. A light juvenile Swainson's Hawk *(Buteo swainsoni)* has conspicuous dark patches on the sides of the neck, buteo proportions, and wingtips nearly to the tailtip.

STATUS: A California Species of Special Concern, and a federal Bird of Conservation Concern.

REMARKS: Prairie Falcons, especially females, have irascible tempers; a trained female whose captured prey has been removed surreptitiously by a falconer may fly into a rage and, suspecting the dog has made off with its quarry, may attack the innocent (and bewildered) canine. Wild Prairies annoyed by the presence of humans near their nest often strike (and sometimes kill outright) accidentally flushed Barn Owls *(Tyto alba)* roosting in the same cliff.

GLOSSARY

Accipiter (ak-ci´-pi-ter) Any of a group of near cosmopolitan woodland hawks in the genus *Accipiter* characterized by short, rounded wings and a long tail.

Aggressive mimicry The imitation of a harmless species or object by a predator to facilitate approaching prey animals.

Albino An individual that appears white because of an inability to produce pigment, the result of receiving recessive genes for that trait from its parents.

Ambush-hunting Remaining concealed on a perch in wait for a prey animal to show itself.

Anthropocentric Interpreting animal behavior in terms of human behavior.

Attacker A raptor that pursues swift, elusive prey.

Axillary A triangle or patch of feathers in a bird's "armpit," where the wing joins the body, or a feather of the axillary.

Bib An area of the upper breast, usually light or surrounded by a crescent of darker feathers.

Billing A mutual rubbing of beaks in birds, usually a component of courtship.

Bloom A grayish coat of fine dust that covers the feathers of some raptors and other birds, especially on the back, produced by specialized feathers, the powderdown.

Brancher A nearly grown juvenile raptor that has left its nest but cannot yet fly.

Breeding bird atlas A book of maps showing the breeding occurrence of birds in a defined area (such as a county).

Brow ridge A small shelf projecting above the eye (eyebrow).

Butcher block An area or object, often a log, branch, or stump, where a raptor plucks prey prior to delivering it to the nest.

Buteo Any of a group of often plump-looking, chiefly rodent-eating hawks, mostly with long, broad wings and short tails.

Caching Hiding killed prey for future use.

Cainism The killing or eviction from the nest of a young raptor by its usually larger or slightly older sibling.

Casting *See* pellet.

Cere A bare, waxy-looking structure at the base of the beak, contiguous with the lips (or gape); it surrounds the nostrils (nares) and is sometimes partly covered with bristles from adjacent skin.

Cerebral hemisphere One of the two halves comprising the brain of vertebrates.

Chaparral Dense vegetation composed of small-leafed shrubs and trees, generally under 4 m (4.3 ft.) in height; highly drought resistant and fire prone.

Clutch The full set of eggs of a bird.

Compromise behavior The behavior of an animal that serves two or more sometimes conflicting purposes; a male raptor's skydance, for example, stimulates a female to breed and simultaneously informs conspecifics that a territory is occupied.

Congener An organism in the same genus as the one under discussion, the term indicating close relationship.

Coniferous forest A forest of cone-bearing trees, usually with a continuous canopy.

Conspecific A bird belonging to the same species as the one under discussion.

Contour-hunting Flying at speed very close to the ground, across and following the folds of the land, in hopes of finding and surprising prey.

Cosmopolitan Occuring worldwide, with few exceptions.

Covert A feather that covers either surface of the wing, usually modified in shape.

Crabbing An aggressive locking of talons by two raptors, often in flight.

Crop A roomy outpocketing of the esophagus at the base of the neck, for temporary storage of ingested food.

Deamination The breakup in the liver of amino acid molecules, one of the steps in the conversion of proteins into sugar (glucose) to provide energy for an animal.

Deciduous woodland Woodland of usually broad-leafed trees that drop their leaves seasonally, with a discontinuous canopy.

Dihedral The position of the wings, in flight, in which they are carried above the horizontal.

Dispersal The movement away from the natal area, generally without returning.

Diurnal Active in the daytime.

Dorsal Pertaining to the back, or upperside, of an animal.

Double-clutching The removal of the first clutch of eggs, thereby inducing the female to lay a second set. This is often done to increase the number of young produced in one season by captive raptors, especially endangered species.

Ecologic formations Distinctive, usually large habitats characterized by certain geologic features or by plant associations that are often linked to elevation.

Ecosystem A major stable system produced by the interactions of organisms between themselves and their nonliving environment.

Ectoparasite A parasite that lives on the surface of the host rather than inside.

End stimulus The ultimate factor that brings on a certain usually innate behavior, such as breeding, after a series of other factors.

Energy pyramid A pyramid composed of organisms, with plants as energy "producers" as the base and meat eaters at the top. It illustrates inefficient energy transfer.

Excreta Droppings and urine.

Exoskeleton A hard, armorlike covering of arthropods to which the muscles are attached from inside and that also serves as a water barrier.

Eyrie A raptor's nest, often used to describe a cliff site.

Falconry The art and sport of training raptors for hunting with humans.

Field mark A distinctive trait that identifies or helps to identify an animal in the wild.

Fledging Leaving the nest (by a young bird).

Flight feathers The large feathers of the wings and tail.

Floater An adult bird capable of breeding but lacking a breeding territory and mate.

Flutter-gliding A slow gliding with the hands of the wing rapidly and shallowly fluttering, as seen, for example, in the American Kestrel *(Falco sparverius)*.

Gape The mouth of the raptor posterior to the beak; also sometimes used to describe the lips, when visible.

Hack The raising and release of young hawks in an artificial nest while providing them with food, eventually allowing them to become independent.

Hand Outer portion of a raptor's wing, holding the primaries and alula.

Holarctic As a noun, the northern hemisphere north of the tropics, including northern Mexico, Mediterranean Africa, and Eurasia; as an adjective, pertaining to that area.

Hovering Remaining stationary in the air by beating the wings, usually rapidly.

Hybrid An offspring resulting from the crossbreeding of two species, showing traits of both.

Immature A nonadult.

Imprint A verb signifying rapid and stable learning early in the life of certain vertebrates that results in recognizing their own species. Many birds hatched in captivity easily misimprint on their human handler. Also, a noun describing a misimprinted raptor.

Incubation The warming a clutch of eggs by sitting or lying on them to promote the development of embryos.

Irruptive Said of species given to sudden, dramatic population increases, frequently on a cyclic basis.

Isotopic analysis The identification of different stable forms of an element, such as lead, to determine its origin, sometimes from minute amounts.

Jesses Leather straps attached to the legs of a falconer's hawk as restraints.

Juvenal An adjective defining plumage or feathering of a bird in its first year (following a coat of down).

Juvenile A noun or adjective describing a young bird in its first year.

Kiting Remaining stationary in the air by the hawk orienting its body into a strong wind.

Kleptoparasite An animal that lives at the expense of another by robbing it.

Kneading A hawk's convulsive clutching of the prey's body in order to kill it, usually alternating the feet.

Leucistic Appearing abnormally colored, the dark browns replaced by a pale coffee and cream color, causing a faded, washed-out appearance.

Local movement The movement of a bird over modest distances, usually away from a breeding area to areas rich in food.

Malar patch; malar stripe A streak or patch extending downward from the lips (gape) or from below the eye, sometimes extending back onto the cheek.

Mantle A triangle of feathers on the upper back between the scapulars.

Mantling The covering of prey by a raptor with spread wings and tail, and simultaneous raising of back and crest feathers.

Mobbing Harassing of a predator (real or perceived, such as a hawk or vulture), by one or more birds, often with the aim of driving it away from a nest site or nesting colony.

Molt The periodic replacement of some or all feathers.

Morph An individual variant of a species, usually designated by color, for example, a rufous morph.

Nape The back of the neck.

Naris A nostril. The plural is *nares*.

Neotropical Pertaining to the tropics of Central and South America.

Nocturnal Active at night.

Nomenclature The bestowing of names.

Occipital crest In birds, a crest of erectile feathers arising from the back of the head (occiput).

Ornithology The study of birds.

Pacific Flyway A path followed by great numbers of migrating birds, chiefly down the western U.S. coast.

Patagium; patagial patch A web of skin at the leading edge of the wing; a patch of dark feathers on its underside, easily seen in flight, is typical in the Red-tailed Hawk *(Buteo jamaicensis).*

Pectoral girdle The assemblage of bones of the forelimbs (wings in birds) and connecting bones to the rib cage and vertebral column.

Pellet An oblong, usually smooth packet of indigestible matter, such as fur, insect parts, accidentally ingested stems and leaves, and sometimes bones, regurgitated by hawks, owls, and some other birds.

Photoperiod The length of day.

Pinions Flight feathers of the wing.

Plumage All feathers of a bird worn at a given time.

Population All members of one species in a given area.

Porro prism binoculars Sometimes bulky binoculars in which the eyepieces are off center from the much wider barrels of the instrument.

Power-diving Aerial diving by raptors, with beating wings (instead of merely falling headfirst) and the beats frequently exaggerated, often during courtship displays.

Power-flying Flying with exaggerated, deep wingbeats, often during courtship display.

Preen To clean, oil, and order the feathers with the beak.

Primaries. The flight feathers of the hand portion of the wing, used for propulsion.

Pull-up The rapid upswing of a raptor following a dive (stoop).

Quartering Flying low to the ground over large areas, usually at moderate speed, to find prey.

Race A variant subpopulation of a species, adapted to a particular environment. If a race is well defined, it is equivalent to a subspecies.

Rectrices The large tail feathers. The singular is *rectrix*.

Remiges The large flight feathers of the wing. The singular is *remex*.

Reversed sexual dimorphism A size phenomenon in some bird species and other animals in which the female is larger than the male.

Rictal bristle A usually very small bristlelike feather that lacks barbs, found near the gape and cere, sometimes covering the nares.

Riparian Pertaining to the banks of bodies of freshwater.

Roof prism binoculars Binoculars in which the eyepieces are centered on the tubular barrels of the instrument.

Rouse To raise and shake the plumage.

Rufous A reddish brown or fox red color.

Savanna A grassland with widely scattered trees.

Scapulars Two tracts of feathers on the back of a bird that cover the wing-body junctions.

Searcher A raptor that pursues relatively easy-to-catch prey, such as grasshoppers or mice.

Sexual dimorphism The visible difference between males and females observed in most vertebrates and frequently expressed in size or color or both.

Short-winged hawk A raptor adapted to life in forests and woodland, with short wings and long tail. In the United States, an accipiter (*see* accipiter).

Shrew A very small insectivorous mammal.

Siblicide *See* cainism.

Skydance An aerial display by raptors associated with courtship.

Soar-hunting Finding prey while soaring.

Soaring Flying with usually fully spread wings and tail, without wingbeats, often in circles or ellipses or along slopes, utilizing thermals and updrafts.

Spatulate Resembling a spatula in shape, like a flattened spoon that widens at the tip.

Spotting scope A monocular telescope popular for studying distant birds, usually mounted on a tripod.

Squeaking Imitating the distress call of a small bird or mammal by "kissing" the back of the hand, to attract a predator.

Still-hunting Hunting from a perch or by sitting on the ground while waiting for prey to show itself.

Stoop The head-long dive, usually at prey, of a raptor, through open air.

Subadult A bird older than a juvenile but younger than an adult, often with distinctive plumage.

Subterminal band A dark horizontal band on the tail short of the tailtip, usually bordered below by a white or pale terminal band (see fig. 10).

Superciliary A white or pale line above the eye along the brow.

Talons The claws.

Tarsi The leg (actually, the foot) segments directly above the toes, usually but not always unfeathered and covered with scales. The singular is *tarsus.*

Telemetry An animal-locating system that comprises a small (often tiny) radio transmitter that can be attached to the animal, and a receiver, equipped with an antenna, to pick up the transmitter signal.

Thermal A bubble or vortex of warm air rising at speed over the ground or human-made substrates heated by the sun.

Thermoregulation The maintenance of a fairly constant body temperature by various means.

Throw-up *See* pull-up.

Trailing edge The rear edge of the spread wing.

Triangulation The process of locating an individual or object by taking bearings from two or more different positions.

Trichomoniasis An infectious, potentially fatal disease caused by a protozoan and often transmitted by pigeons and doves.

Underparts The underside of a bird, including the breast, belly, and undertail coverts.

Undulating Moving up and down, like waves in a row.

Upperparts The top of a bird, including the back, rump, and scapulars.

Urates The nitrogenous wastes from the kidneys, produced by the breakdown of protein.

Urohydrosis The voiding of excreta onto the legs to cool off, practiced by New World vultures and some other birds.

Ventral Pertaining to the underside.

Vole Any of several species of mouse-sized, usually short-tailed, rodents, some of which are important raptor food.

West Nile Virus A viral disease, carried chiefly by migratory birds and transmitted by the bites of mosquitos and perhaps louse flies (Hippoboscidae) or the ingestion of infected prey. Bird species most affected include crows and their allies, owls, and most diurnal raptors.

Whitewash Raptor droppings, often conspicuous on cliffs because of the chalk-white color of the uric acid contents.

Wing flex A quick flap downward, chiefly at the hand joint, so that when seen head-on, the bird forms a rectangle with its wings; the wings are then returned to the normal gliding position. Typical of the three vultures found in the United States.

Wing lining The underwing coverts.

Wing-loading The amount of weight that a unit area of wing must support.

Wing window A conspicuous light area in the hand portion of the spread wing.

REFERENCES
AND FURTHER READING

Balgooyen, T. G. 1988. A unique encounter among a Gyrfalcon, Peregrine Falcon, Prairie Falcon and American Kestrel. *Journal of Raptor Research* 22:71

Bednarz, J. C. 1987. Pair and group reproductive success, polyandry, and cooperative breeding in Harris' Hawk. *Auk* 104:393–404.

Bell, D. A., D. P. Gregoire, and B. J. Walton. 1996. Bridge use by Peregrine Falcons in the San Francisco Bay Area. In *Raptors in Human Landscapes,* eds. D. M. Bird, E. E. Varland, and J. J. Negro, 15–24. New York: Academic Press.

Bent, A. C. 1937. *Life histories of North American birds of prey, parts 1 and II.* New York: Dover Publications.

Bildstein, K. L., and K. Meyer. 2000. Sharp-shinned Hawk *(Accipiter striatus).* In *The Birds of North America, No. 482,* eds. A. Poole and F. Gill. Philadelphia, PA: The Birds of North America.

Bloom, P. H. 1985. Raptor movements in California. In *Proceedings of Hawk Migration Conference IV,* ed. M. Harwood, 313–324. North Wales, PA: Hawk Migration Association of North America.

Bloom, P. H. 1980. *The status of the Swainson's Hawk in California, 1979.* Final Report II-8.0, Bureau of Land Management and Federal Aid in Wildlife Restoration, Project W-54-R-12, Calif. Dept. of Fish and Game. Sacramento: The Resources Agency, California Department of Fish and Game, and United States Bureau of Land Management.

Boal, C. W. 2001. Agonistic behavior of Cooper's Hawks. *Journal of Raptor Research* 35:253–256.

Brown, L., and D. Amadon. 1968. *Eagles, hawks and falcons of the world.* New York: McGraw-Hill.

Brown, N. L. 1996. Swainson's Hawk profile. Endangered Species Re-

covery Program, California State University Stanislaus. http://
esrpweb.csustan.edu/speciesprofiles/profile.php?sp = busw. Accessed 16 Feb. 2004.

Cade, T.J., and W. Burnham, eds. 2003. *Return of the Peregrine: A
North American saga of tenacity and teamwork.* Boise, ID: The
Peregrine Fund.

Cade, T.J., J.H. Enderson, C.G. Thelander, and C.M. White, eds.
1988. *Peregrine falcon populations.* Boise, ID: The Peregrine Fund.

California Department of Fish and Game. 2003. Bird species of special concern. www.dfg.ca.gov/hcpb/species/ssc/sscbird/sscbird.
shtml. Accessed 25 April 2004.

California Department of Fish and Game. 2003. Threatened and endangered birds. www.dfg.ca.gov/hcpb/species/t_e_spp/tebird/
tebirda.shtml. Accessed 25 April 2004.

Clark, W.S., and B.K. Wheeler. 2001. *A field guide to hawks of North
America.* Boston: Houghton Mifflin.

Dawson, J.W., and R.W. Mannan. 1991a. Dominance hierarchies and
helper contributions in Harris's Hawks. *Auk* 108:649–660.

Dawson, J.W., and R.W. Mannan. 1991b. The role of territoriality in
the social organization of Harris's Hawks. *Auk* 108:661–672.

Dechant, J.A., M.L. Sondreal, D.H. Johnson, L.D. Igl, C.M. Goldade,
A.L. Zimmerman, and B.R. Euliss. 2003. Effects of management
practices on grassland birds: Ferruginous Hawk. Northern
Prairie Wildlife Research Center, Jamestown, ND. www.npwrc
.usgs.gov/resource/literatr/grasbird/feha/feha.htm (Version 12
DEC2003). Accessed 14 Jan. 2004.

DiDonato, J.E. 1992. Intraspecific nest defense by Prairie Falcons.
Journal of Raptor Research 26:40.

England, A.S., J.A. Estep, and W.R. Holt. 1995. Nest-site selection
and reproductive performance of urban-nesting Swainson's
Hawks in the Central Valley of California. *Journal of Raptor Research* 29:179–186.

Erichsen, A.L., A.M. Commandatore, and D.M. Fry. 1995. Communal roosts: Seasonal dynamics of a White-tailed Kite population
in the Sacramento Valley, California. *Journal of Raptor Research*
29:70.

Erickson, M.M. 1937. A jay shoot in California. *Condor* 39:111–115.

Estep, J.A., and S. Teresa. 1992. Regional conservation planning for
the Swainson's Hawk *(Buteo swainsoni)* in the Central Valley of
California. In *Wildlife 2001: Populations,* eds. D.R. McCullough
and R.H. Barrett, 775–789. New York: Elsevier Applied Science.

Ferguson-Lees, J., and D.A. Christie. 2001. *Raptors of the world.* Boston: Houghton Mifflin.

Fish, A. 2001. *Season summary.* San Francisco: Golden Gate Raptor Observatory.

Fish, A. 1999. She taps on the roof; she does a little dance: A conversation with a woman who feeds a hawk. *Pacific Raptor Report* 20:8–10.

Fox, N. 1995. *Understanding the bird of prey.* Blaine, WA: Hancock House Publishers.

Franklin, K. 1999. Vertical flight. *North American Falconers' Association Journal.* 38:68–72.

Garcelon, D.K., R.W. Risebrough, W.M. Jarman, A.B. Chartrand, and E.E. Littrell. 1989. Accumulation of DDE by Bald Eagles reintroduced to Santa Catalina Island in southern California. In *Proceedings of the Third World Conference on Birds of Prey and Owls,* eds. B-U. Meyburg and R.D. Chancellor, 491–494. Berlin: The World Working Group on Birds of Prey and Owls.

Garrison, B.A., and P.H. Bloom. 1993. Natal origins and winter site fidelity of Rough-legged Hawks wintering in California. *Journal of Raptor Research* 27:116–118.

Gessaman, J.A. 1980. An evaluation of heart rate as an indirect measure of daily energy metabolism of the American Kestrel. *Comparative Biochemistry and Physiology.* 65A:273–289.

Golet, G.H., H.T. Golet, and A.M. Colton. 2003. Immature Northern Goshawk captures, kills, and feeds on adult-sized Wild Turkey. *Journal of Raptor Research* 37:337–340.

Grenfell, W.E., Jr., and W.F. Laudenslayer Jr., eds. 1983. *The distribution of California birds.* California Wildlife/Habitat Relationships Program, Publication No. 4. Sacramento: California Department of Fish and Game; San Francisco: United States Forest Service.

Griffiths, C.S. 1994. Monophyly of the falconiformes based on syringeal morphology. *Auk* 111:787–805.

Grinnell, J., and A.H. Miller. 1944. *The distribution of the birds of California.* Berkeley: Cooper Ornithological Club.

Haak, B.A. 1995. *Pirate of the plains.* Blaine, WA: Hancock House Publishers.

Hall, L.S., A.M. Fish, and M.L. Morrison. 1992. The influence of weather on hawk movements in coastal northern California. *Wilson Bulletin* 104:447–461.

Hamerstrom, F., F.N. Hamerstrom, and D.J. Burke. 1985. Effect of voles on mating systems in a central Wisconsin population of harriers. *Wilson Bulletin* 97:332–346.

Harvey, P.H., P.J. Greenwood, and C.M. Perrins. 1979. Breeding area fidelity of Great Tits *(Parus major)*. *Journal of Animal Ecology.* 48:305–313.

Heath, J.E. 1962. Temperature fluctuation in the Turkey Vulture. *Condor* 64:234–235.

Heizer, R.F., ed. 1978. *California.* Vol. 8 of *Handbook of North American Indians.* Washington, DC: Smithsonian Institution.

Henny, C.J., and R.G. Anthony. 1989. Bald eagle and Osprey. In *Proceedings of the Western Raptor Management Symposium and Workshop,* ed. B.G. Pendleton, 66–82. Washington, DC: National Wildlife Federation.

Herman, S.G., M.N. Kirven, and R.W. Risebrough. 1970. The Peregrine Falcon decline in California. *Audubon Field Notes* 24: 609–613.

Hoffman, S.W., J.P. Smith, and T.D. Meehan. 2002. Breeding grounds, winter ranges, and migratory routes of raptors in the mountain west. *Journal of Raptor Research* 36:97–110.

Houghton, L.M., and L.M. Rymon. 1997. Nesting distribution and population status of U.S. Ospreys 1994. *Journal of Raptor Research* 31:44–53.

Hunt, W.G., R.E. Jackman, J.M. Jenkins, C.G. Thelander, and R.N. Lehman. 1992. Northward post-fledging migration of California Bald Eagles. *Journal of Raptor Research* 26:19–23.

Hunt, W.G., R.E. Jackman, T.L. Hunt, D.E. Driscoll, and L. Culp. 1998. A population study of Golden Eagles in the Altamont Pass Wind Resource Area: Population trend analysis 1997. Report to National Renewable Energy Laboratory, Subcontract XAT-6-16459-01. Santa Cruz: Predatory Bird Research Group, University of California.

Hunting, K.W. 1998. Ferruginous Hawk species account. In *California partners in flight grassland bird conservation plan,* ed. B. Allen. California Partners in Flight and Point Reyes Bird Observatory. www.prbo.org/calpif/htmldocs/grassland.html. Accessed 5 March 2004.

Institute for Wildlife Studies. 2000. Osprey restoration. www.iws.org/osprey2.htm. Accessed 14 June 2003.

Jenkins, J.M. 1992. Ecology and behavior of a resident population of Bald Eagles. PhD diss. University of California, Davis.

Jenkins, J.M., R.E. Jackman, and W.G. Hunt. 1999. Survival and movements of immature Bald Eagles fledged in northern California. *Journal of Raptor Research* 33:81–86.

Jesus, D. F., L. J. Jesus, A. C. Hull, A. M. Fish, and A. K. Francone. 1995. When juveniles look like adults: Gray Cooper's Hawks in the San Francisco Bay Area. *Journal of Raptor Research* 29:70–71.

Johnsgard, P. A. 1990. *Hawks, eagles, and falcons of North America.* Washington, DC: Smithsonian Institution Press.

Kerlinger, P. 1989. *Flight strategies of migrating hawks.* Chicago: University of Chicago Press.

Kiff, L. F. 1980. Historical changes in resident populations of California islands raptors. In *The California islands: Proceedings of a multidisciplinary symposium,* ed. D. M. Power, 651–673. Santa Barbara: Santa Barbara Museum of Natural History.

Koford, C. B. 1953. *The California Condor.* National Audubon Society Research Report No. 4:1–154.

Koivula, M., and J. Viitala. 1999. Rough-legged Buzzards use vole scent marks to assess hunting areas. *Journal of Avian Biology* 30:329–332.

Kroeber, A. L. 1953. *Handbook of the Indians of California.* Berkeley: California Book.

Kurosawa, T., and R. Kurosawa. 2003. A helper at the nest of Peregrine Falcons in northern Japan. *Journal of Raptor Research* 37:340–342.

Mallette, R. D., and G. I. Gould Jr. 1976. *Raptors of California.* Sacramento: California Department of Fish and Game.

McCaskie, G., and M. San Miguel. 1999. Report of the California Bird Records Committee: 1996 records. *Western Birds* 30:57–85.

Millar, J. G. 2002. The protection of eagles and the Bald and Golden Eagle Protection Act. *Journal of Raptor Research* 36(1, Suppl.):29–31.

Miller, A. H. 1951. *An analysis of the distribution of the birds of California.* Berkeley: University of California Press.

Mueller, H. C. 1972. Zone-tailed Hawk and Turkey Vulture: Mimicry or aerodynamics? *Condor* 74:221–222.

Newton, I. 1986. *The Sparrowhawk.* Calton, England: T. & A. D. Poyser.

Olendorff, R. R., D. D. Bibles, M. T. Dean, J. R. Haugh, and M. N. Kochert. 1989. Raptor habitat management under the U.S. Bureau of Land Management multiple-use mandate. *Raptor Research Reports* 8:1–80.

Oliphant, L. W. 1991. Hybridization between a Peregrine Falcon and a Prairie Falcon in the wild. *Journal of Raptor Research* 25:36–39.

Olsen, J. 1986. An unusual incident with the Bald Eagle. *Journal of Raptor Research* 20:41.

Palmer, R. S., ed. 1988. *Handbook of North American birds, Vols. 4 and 5.* New Haven: Yale University Press.

Patten, M. A. and R. A. Erickson. 2000. Population fluctuations of the Harris' Hawk and its reappearance in California. *Journal of Raptor Research* 34:187–195.

Peeters, H. J. 1994. Suspected poisoning of Golden Eagles *Aquila chrysaetos* by chlorophacinone. In *Raptor Conservation Today,* eds. B.-U. Meyburg, and R. D. Chancellor, 775–776. Berlin: World Working Group on Birds of Prey, The Pica Press.

Peeters, H. J. 1963a. Einiges über den Waldfalken *Micrastur semitorquatus. Journal für Ornithologie* 104:357–364.

Peeters, H. J. 1963b. Two observations of avian predation. *Wilson Bulletin* 75:274.

Prum, R. O., and A. H. Brush. 2003. Which came first, the feather or the bird? *Scientific American* 288(3):84–93.

Remple, J. D. 2002. Understanding the avian respiratory system: Facts and considerations for the falconer. *North American Falconers' Association Journal* 41:72–75.

Risebrough, R. W., R. W. Schlorff, P. H. Bloom, and E. E. Littrell. 1989. Investigations of the decline of Swainson's Hawk populations in California. *Journal of Raptor Research* 23:63–71.

Rottenborn, S. C. 2000. Nest-site selection and reproductive success of urban Red-Shouldered Hawks in central California. *Journal of Raptor Research* 34:18–25.

Rouan, L. 1996. The hawk is where? The 1995 telemetry season. *Pacific Raptor Report* 17:4–6.

Sauer, J. R., J. E. Hines, and J. Fallon. 2003. *The North American breeding bird survey, results and analysis 1966–2000.* Version 2003.1. Laurel, MD: USGS Patuxent Wildlife Research Center.

Scheuermann, K. 1996. GGRO band recoveries 1992–1996. *Pacific Raptor Report* 17:11–16.

Schlorff, R. W. and P. H. Bloom. 1983. Importance of riparian systems to nesting Swainson's Hawks in the Central Valley of California. In *California riparian systems,* eds. R. E. Warner and K. M. Hendrix, 612–618. Berkeley: University of California Press.

Sharp, C. S. 1902. Nesting of Swainson's Hawk. *Condor* 4:116–118.

Sibley, C., and J. Ahlquist. 1990. *Phylogeny and classification of birds of the world.* New Haven: Yale University Press.

Small, A. 1994. *California birds: Their status and distribution.* Vista, CA: Ibis Publishing.

Smith, H.R., R.M. DeGraaf, and R.S. Miller. 2002. Exhumation of food by Turkey Vulture. *Journal of Raptor Research* 36:144–145.

Snyder, N.F.R., and J.A. Hamber. 1985. Replacement clutching and annual nesting of California Condors. *Condor* 87:374–378.

Snyder, N.F.R., and N.J. Schmitt. 2002. California Condor *(Gymnogyps californianus)*. In *The birds of North America, No. 610,* ed. A. Poole and F. Gill. Philadelphia: The Birds of North America.

Snyder, N.F.R., and H. Snyder. 2000. *The California Condor, a saga of natural history and conservation.* London: Academic Press.

Snyder, N.F.R., and H. Snyder. 1991. *Raptors: North American birds of prey.* Stillwater, MN: Voyageur Press.

Sodhi, N.S., L.W. Oliphant, P.C. James, and I.G. Warkentin. 1993. Merlin *(Falco columbarius)*. In *The Birds of North America, No. 44,* eds. A. Poole and F. Gill. Philadelphia: Academy of Natural Sciences; Washington, DC: American Ornithologists' Union.

Sprunt, A. 1955. *North American birds of prey.* New York: Harper & Brothers.

Steenhof, K. 1998. Prairie Falcon *(Falco mexicanus)*. In *The Birds of North America, No. 346,* eds. A. Poole and F. Gill. Philadelphia: The Birds of North America.

Stefanek, P.R., W.W. Bowerman, T.G. Grubb, and J.B. Holt. 1992. Nestling Red-tailed Hawk in occupied Bald Eagle nest. *Journal of Raptor Research* 26:40–41.

Steinhart, P. 1990. *California's wild heritage: Threatened and endangered animals in the golden state.* Sacramento: California Department of Fish and Game; San Francisco: California Academy of Sciences and Sierra Club Books.

Timbrook, J., and J.R. Johnson. 1999. People of the sky: Birds in Chumash culture. Paper presented at the 22nd Ethnobiology Conference, Oaxaca, Mexico. www.sbnature.org/research/anthro/chbirds.htm. Accessed 2 Jan. 2004.

Tlusty, M.F., and F. Hamerstrom. 1992. American Kestrels *(Falco sparverius)* adopt and fledge European Starlings *(Sturnus vulgaris)*. *Journal of Raptor Research* 26:195.

Tucker, V.A. 1993. Gliding birds: Reduction of induced drag by wing tip slots between the primary feathers. *Journal of Experimental Biology.* 180:285–310.

Tucker, V.A. 1992. Pitching equilibrium, wing span and tail span in a gliding Harris' Hawk, *Parabuteo unicinctus. Journal of Experimental Biology.* 165:21–41.

Tucker, V. A. 1991. The effect of molting on the gliding performance of a Harris' Hawk *(Parabuteo unicinctus)*. *Auk* 108:108–113.

U.S. Department of the Interior, Fish and Wildlife Service. 1983. Permits to take Golden Eagle nests. Rules and regulations. *Federal Register* 48(251):57295–57301.

Viitala, J., E. Korpimäki, P. Palokangas, and M. Koivula. 1995. Attraction of kestrels to vole scent marks visible in ultraviolet light. *Nature* 373:425–427.

Villarroel, M. R., and D. M. Bird. 1994. Are American Kestrels promiscuous nymphomaniacs? *Journal of Raptor Research* 28:66.

Wakeley, J. S. 1978. Factors affecting the use of hunting sites by Ferruginous Hawks. *Condor* 80:316–326.

Walton, B. J. 2003. Restoration of the Peregrine population in California. In *Return of the Peregrine,* ed. T. J. Cade and W. Burnham, 155–171. Boise, ID: The Peregrine Fund.

Walton, B. J. 1978. Peregrine–Prairie Falcon interaction. *Journal of Raptor Research* 12:46–47.

Weidensaul, S. 2000. *The raptor almanac.* New York: The Lyons Press.

Welty, J. C. 1975. *The life of birds.* Philadelphia: W. B. Saunders.

Wendt, A., G. Septon, and J. Moline. 1991. Juvenile urban-hacked Peregrine Falcons hunt at night. *Journal of Raptor Research* 25:94–95.

Wheeler, B. K. 2003. *Raptors of western North America.* Princeton: Princeton University Press.

White, C. M., N. J. Clum, T. J. Cade, and W. G. Hunt. 2002. Peregrine Falcon *(Falco peregrinus).* In *The Birds of North America, No. 660,* eds. A. Poole and F. Gill. Philadelphia: The Birds of North America.

Williams, T. 2000. Zapped! *Audubon Magazine* 102.1:32–44.

Woodbridge, B., K. K. Finley, and P. H. Bloom. 1995. Reproductive performance, age structure, and natal dispersal of Swainson's Hawks in the Butte Valley, California. *Journal of Raptor Research* 29:187–192.

Woodbridge, B., K. K. Finley, and S. T. Seager. 1995. An Investigation of the Swainson's Hawk in Argentina. *Journal of Raptor Research* 29:202–204.

ADDITIONAL CAPTIONS

PAGES II–III Red-shouldered Hawk perched in a red alder *(Alnus rubra)*.

PAGE VI Big Sur Peregrine Falcon.

PAGES XIV–1 Peregrine Falcon stooping at American Wigeon *(Anas americana)*.

PAGES 46–47 Golden Eagles chasing California Jackrabbit *(Lepus californicus)*.

PAGES 76–77 Northern Goshawk in coniferous forest.

PAGES 106–107 American Kestrel perched on rock with petroglyphs.

PAGES 136–137 Sharp-shinned Hawk chasing Dark-eyed Juncos *(Junco hyemalis)*.

INDEX

Page numbers in **bold** indicate main discussion of taxon.

ABOUT THE AUTHORS

Hans Peeters is an ornithologist whose main interest is raptors. Born in Germany in 1937, he came to the United States at the age of 16, received a B.A. in Comparative Literature at the University of California, Berkeley, and went on to complete a graduate degree in zoology at the same institution. For 37 years he taught subjects ranging from ecology to zoology and field biology at Chabot College in Hayward, California. In addition to authoring several scientific papers, he is a coauthor, with E.W. Jameson Jr., of *Mammals of California* (University of California Press) and of a second book about falconry. In the past 20 years, his avocation of painting has developed into a second career, and he much enjoys combining his interests in science and art wherever possible. In recent years, he has been painting endangered species for conservation postage stamps for Mexico, and other conservation work includes painting a poster of a Harpy Eagle for the country of Panama, where this national bird is endangered. Hans and Pam, his wife and partner of 30 years, make their home in Sunol, California, where from their kitchen window they see as many as six different raptor species throughout the year. Their acre and a half of oak-bay and riparian woodland property holds 10 species of snakes, a delight for a zoologist.

Pam Peeters, a native Californian, attended the University of California at Berkeley and is herself a published author. Together with her husband, she has watched raptors over almost all the world. She is also very interested in plants and is responsible for teaching her husband that these organisms are nearly as interesting as animals.

Series Design:	Barbara Jellow
Design Enhancements:	Beth Hansen
Design Development:	Jane Tenenbaum
Cartographer:	Bill Nelson
Composition:	Impressions Book and Journal Services, Inc.
Text:	9/10.5 Minion
Display:	Franklin Gothic Book and Demi
Printer and binder:	Everbest Printing Company